P9-CCO-970

MY BEST MATHEMATICAL AND LOGIC PUZZLES

Martin Gardner

DOVER PUBLICATIONS, INC.
New York

Copyright

Copyright © 1994 by Martin Gardner.
All rights reserved.

Bibliographical Note

My Best Mathematical and Logic Puzzles, first published by Dover Publications, Inc., in 1994, is a new and original collection of work previously published in *Scientific American, Games* magazine and in earlier volume collections. A new Introduction has been written expressly for this Dover edition.

Library of Congress Cataloging-in-Publication Data

Gardner, Martin, 1914–
My best mathematical and logic puzzles / Martin Gardner.
p. cm.
ISBN-13: 978-0-486-28152-0 (pbk.)
ISBN-10: 0-486-28152-3 (pbk.)
1. Mathematical recreations. I. Title.
QA95.G292 1994
793.7′4—dc20 94-25660

Manufactured in the United States by LSC Communications
28152319 2017
www.doverpublications.com

INTRODUCTION

DURING the twenty-five years that I had the great privilege of writing the Mathematical Games column in *Scientific American*, my practice was to devote a column, about every six months, to what I called short problems or puzzles. These puzzles were, of course, mathematical rather than problems involving words. I did my best to present new and unfamiliar puzzles that were not to be found in classic collections such as the books by Sam Loyd and Henry Dudeney.

Readers were quick to catch mistakes and to supply in some cases alternate solutions or interesting generalizations. This valuable feedback was incorporated when the puzzle columns were reprinted in book collections.

Most of the problems in this book are selected from the first three collections. The last 12 puzzles are selected from two articles I contributed to *Games* magazine (January/February and November/December, 1978). Some have been updated by adding references to new developments related to the puzzle. Needless to say, I welcome any corrections or additions. They can be sent to me in care of the publisher, Dover Publications, 31 East 2nd Street, Mineola, N.Y. 11501.

MARTIN GARDNER

CONTENTS

PUZZLES

1. The Returning Explorer

AN OLD RIDDLE runs as follows. An explorer walks one mile due south, turns and walks one mile due east, turns again and walks one mile due north. He finds himself back where he started. He shoots a bear. What color is the bear? The time-honored answer is: "White," because the explorer must have started at the North Pole. But not long ago someone made the discovery that the North Pole is not the only starting point that satisfies the given conditions! Can you think of any other spot on the globe from which one could walk a mile south, a mile east, a mile north and find himself back at his original location?

2. Draw Poker

TWO MEN PLAY a game of draw poker in the following curious manner. They spread a deck of 52 cards face up on the table so that they can see all the cards. The first player draws a hand by picking any five cards he chooses. The second player does the same. The first player now may keep his original hand or draw up to five cards. His discards are put aside out of the game. The second player may now draw likewise. The person with the higher hand then wins. Suits have equal value, so that two flushes tie unless one is made of higher cards. After a while the players discover that the first player can always win if he draws his first hand correctly. What hand must this be?

3. The Mutilated Chessboard

THE PROPS FOR this problem are a chessboard and 32 dominoes. Each domino is of such size that it exactly covers two adjacent squares on the board. The 32 dominoes therefore can cover all 64 of the chessboard squares. But now suppose we cut off two squares at diagonally opposite corners of the board and discard one of the dominoes. Is it possible to place the 31 dominoes on the board so that all the remaining 62 squares are covered? If so, show how it can be done. If not, prove it impossible.

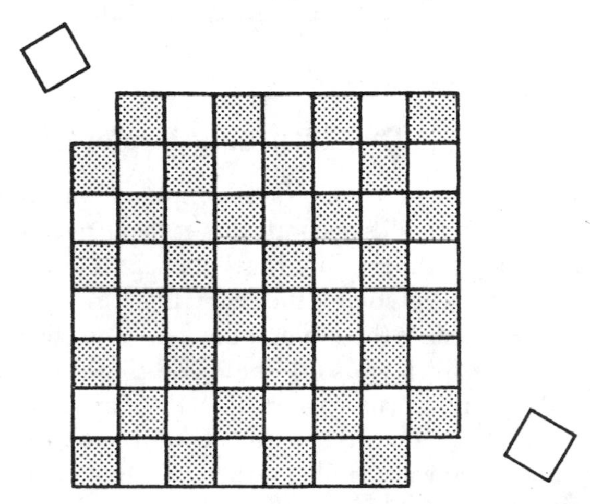

4. The Fork in the Road

HERE'S A RECENT twist on an old type of logic puzzle. A logician vacationing in the South Seas finds himself on an island inhabited by the two proverbial tribes of liars and truth-tellers. Members of one tribe always tell the truth, members of the other always lie. He comes to a fork in a road and has to ask a native bystander which branch he should take to reach a village. He has no way of telling whether the native is a truth-teller or a liar. The logician thinks a moment, then asks *one* question only. From the reply he knows which road to take. What question does he ask?

5. Scrambled Box Tops

IMAGINE THAT YOU have three boxes, one containing two black marbles, one containing two white marbles, and the third, one black marble and one white marble. The boxes were labeled for their contents—BB, WW and BW—but someone has switched the labels so that every box is now incorrectly labeled. You are allowed to take one marble at a time out of any box, without looking inside, and by this process of sampling you are to determine the contents of all three boxes. What is the smallest number of drawings needed to do this?

6. Cutting the Cube

A CARPENTER, working with a buzz saw, wishes to cut a wooden cube, three inches on a side, into 27 one-inch cubes. He can do this easily by making six cuts through the cube, keeping the pieces together in the cube shape. Can he reduce the number of necessary cuts by rearranging the pieces after each cut?

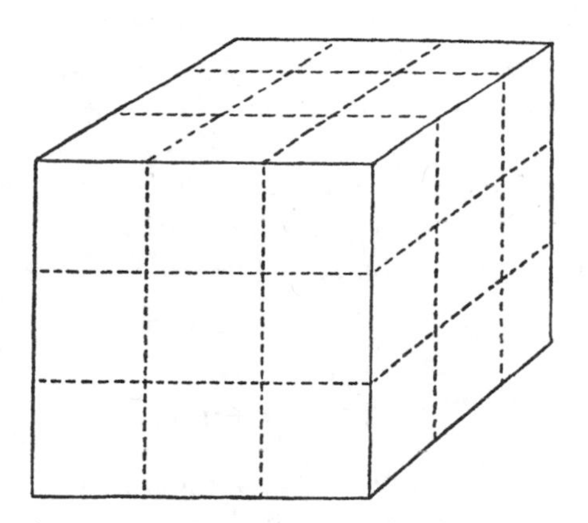

7. Bronx vs. Brooklyn

A YOUNG MAN lives in Manhattan near a subway express station. He has two girl friends, one in Brooklyn, one in The Bronx. To visit the girl in Brooklyn he takes a train on the downtown side of the platform; to visit the girl in The Bronx he takes a train on the uptown side of the same platform. Since he likes both girls equally well, he simply takes the first train that comes along. In this way he lets chance determine whether he rides to The Bronx or to Brooklyn. The young man reaches the subway platform at a random moment each Saturday afternoon. Brooklyn and Bronx trains arrive at the station equally often—every 10 minutes. Yet for some obscure reason he finds himself spending most of his time with the girl in Brooklyn: in fact on the average he goes there nine times out of ten. Can you think of a good reason why the odds so heavily favor Brooklyn?

8. The Early Commuter

A COMMUTER IS in the habit of arriving at his suburban station each evening exactly at five o'clock. His wife always meets the train and drives him home. One day he takes an earlier train, arriving at the station at four. The weather is pleasant, so instead of telephoning home he starts walking along the route always taken by his wife. They meet somewhere on the way. He gets into the car and they drive home, arriving at their house ten minutes earlier than usual. Assuming that the wife always drives at a constant speed, and that on this occasion she left just in time to meet the five o'clock train, can you determine how long the husband walked before he was picked up?

9. The Counterfeit Coins

IN RECENT YEARS a number of clever coin-weighing or ball-weighing problems have aroused widespread interest. Here is a new and charmingly simple variation. You have 10 stacks of coins, each consisting of 10 half-dollars. One entire stack is counterfeit, but you do not know which one. You do know the weight of a genuine half-dollar and you are also told that each counterfeit coin weighs one gram more than it should. You may weigh the coins on a pointer scale. What is the smallest number of weighings necessary to determine which stack is counterfeit?

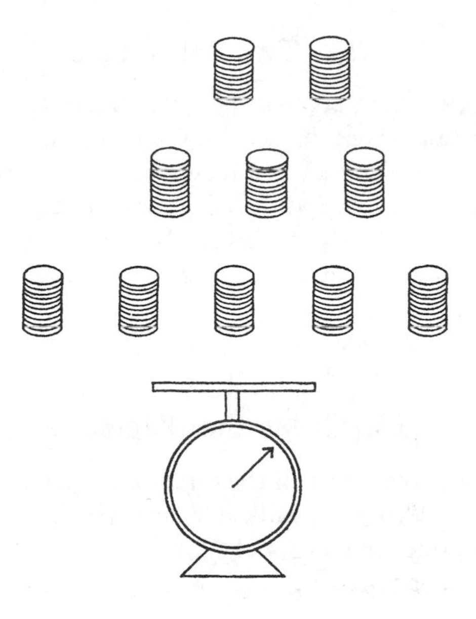

10. The Touching Cigarettes

FOUR GOLF BALLS can be placed so that each ball touches the other three. Five half-dollars can be arranged so that each coin touches the other four.

Is it possible to place six cigarettes so that each touches the other five? The cigarettes must not be bent or broken.

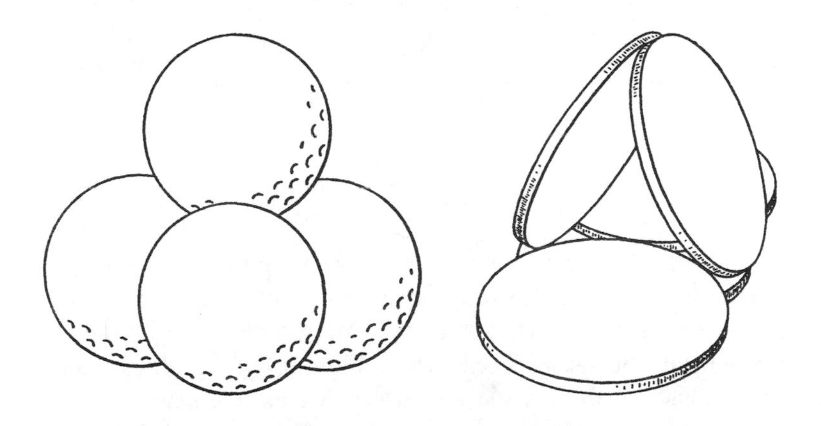

11. Two Ferryboats

TWO FERRYBOATS start at the same instant from opposite sides of a river, traveling across the water on routes at right angles to the shores. Each travels at a constant speed, but one is faster than the other. They pass at a point 720 yards from the nearest shore. Both boats remain in their slips for 10 minutes before starting back. On the return trips they meet 400 yards from the other shore.

How wide is the river?

12. Guess the Diagonal

A RECTANGLE is inscribed in the quadrant of a circle as shown. Given the unit distances indicated, can you accurately determine the length of the diagonal AC?

Time limit: one minute!

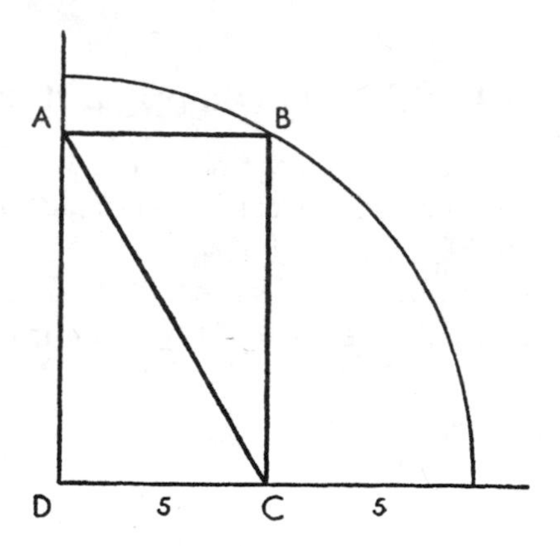

13. Cross the Network

ONE OF THE oldest of topological puzzles, familiar to many a schoolboy, consists of drawing a continuous line across the closed network shown so that the line crosses each of the 16 segments of the network only once. The curved line shown here does not solve the puzzle because it leaves one segment uncrossed. No "trick" solutions are allowed, such as passing the line through a vertex or along one of the segments, folding the paper and so on.

It is not difficult to prove that the puzzle cannot be solved on a plane surface. Two questions: Can it be solved on the surface of a sphere? On the surface of a torus (doughnut)?

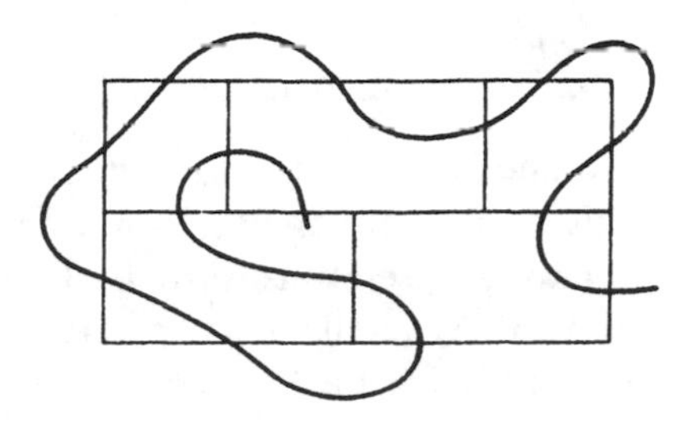

14. The 12 Matches

ASSUMING that a match is a unit of length, it is possible to place 12 matches on a plane in various ways to form polygons with integral areas. The illustration shows two such polygons: a square with an area of nine square units, and a cross with an area of five.

The problem is this: Use all 12 matches (the entire length of each match must be used) to form in similar fashion the perimeter of a polygon with an area of exactly four square units.

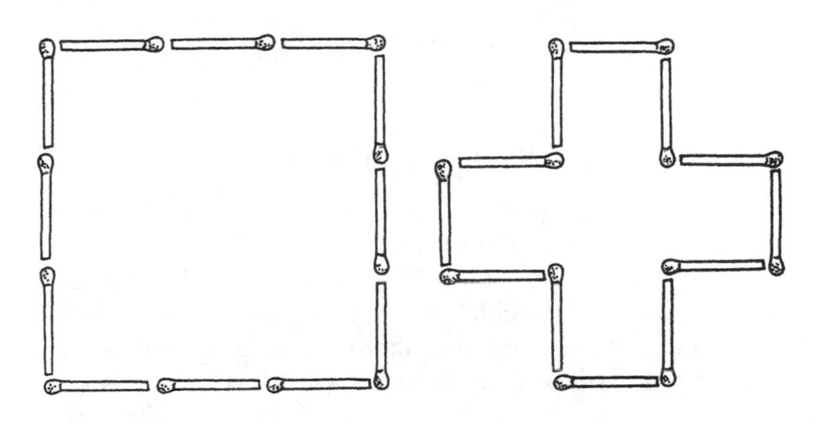

15. Hole in the Sphere

THIS IS an incredible problem—incredible because it seems to lack sufficient data for a solution. A cylindrical hole six inches long has been drilled straight through the center of a solid sphere. What is the volume remaining in the sphere?

16. The Amorous Bugs

FOUR BUGS—A, B, C and D—occupy the corners of a square 10 inches on a side. A and C are male, B and D are female. Simultaneously A crawls directly toward B, B toward C, C toward D and D toward A. If all four bugs crawl at the same constant rate, they will describe four congruent logarithmic spirals which meet at the center of the square.

How far does each bug travel before they meet? The problem can be solved without calculus.

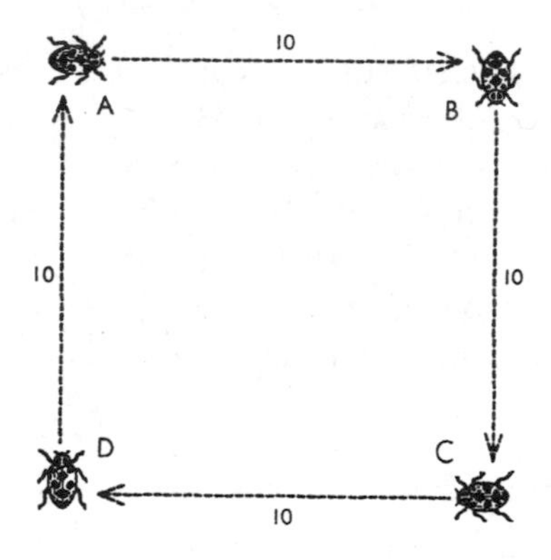

17. How Many Children?

"I HEAR some youngsters playing in the backyard," said Jones, a graduate student in mathematics. "Are they all yours?"

"Heavens, no," exclaimed Professor Smith, the eminent number theorist. "My children are playing with friends from three other families in the neighborhood, although our family happens to be largest. The Browns have a smaller number of children, the Greens have a still smaller number, and the Blacks the smallest of all."

"How many children are there altogether?" asked Jones.

"Let me put it this way," said Smith. "There are fewer than 18 children, and the product of the numbers in the four families happens to be my house number which you saw when you arrived."

Jones took a notebook and pencil from his pocket and started scribbling. A moment later he looked up and said, "I need more information. Is there more than one child in the Black family?"

As soon as Smith replied, Jones smiled and correctly stated the number of children in each family.

Knowing the house number and whether the Blacks had more than one child, Jones found the problem trivial. It is a remarkable fact, however, that the number of children in each family can be determined solely on the basis of the information given above!

18. The Twiddled Bolts

TWO IDENTICAL BOLTS are placed together so that their helical grooves intermesh. If you move the bolts around each other as you would twiddle your thumbs, holding each bolt firmly by the head so that it does not rotate and twiddling them in the direction shown, will the heads (a) move inward, (b) move outward, or (c) remain the same distance from each other? The problem should be solved without resorting to actual test.

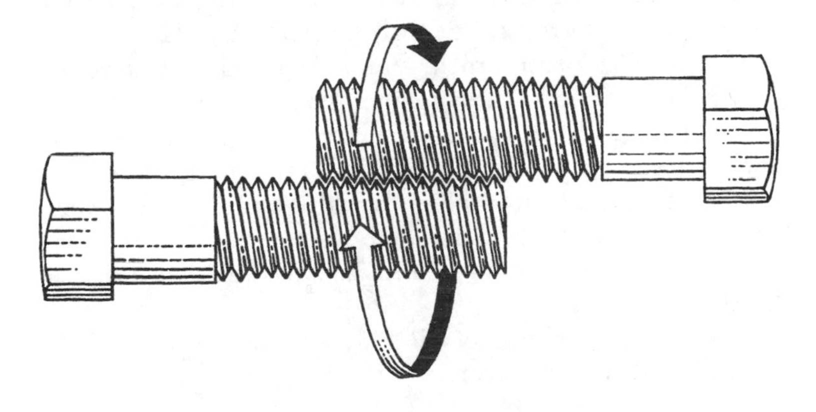

19. The Flight around the World

A GROUP of airplanes is based on a small island. The tank of each plane holds just enough fuel to take it halfway around the world. Any desired amount of fuel can be transferred from the tank of one plane to the tank of another while the planes are in flight. The only source of fuel is on the island, and for the purposes of the problem it is assumed that there is no time lost in refueling either in the air or on the ground.

What is the smallest number of planes that will ensure the flight of one plane around the world on a great circle, assuming that the planes have the same constant ground speed and rate of fuel consumption and that all planes return safely to their island base?

20. The Repetitious Number

AN UNUSUAL parlor trick is performed as follows. Ask spectator A to jot down any three-digit number, and then to repcat the digits in the same order to make a six-digit number (*e.g.*, 394,394). With your back turned so that you cannot see the number, ask A to pass the sheet of paper to spectator B, who is requested to divide the number by 7.

"Don't worry about the remainder," you tell him, "because there won't be any." B is surprised to discover that you are right (*e.g.*, 394,394 divided by 7 is 56,342). Without telling you the result, he passes it on to spectator C, who is told to divide it by 11. Once again you state that there will be no remainder, and this also proves correct (56,342 divided by 11 is 5,122).

With your back still turned, and no knowledge whatever of the figures obtained by these computations, you direct a fourth spectator, D, to divide the last result by 13. Again the division comes out even (5,122 divided by 13 is 394). This final result is written on a slip of paper which is folded and handed to you. Without opening it you pass it on to spectator A.

"Open this," you tell him, "and you will find your original three-digit number."

Prove that the trick cannot fail to work regardless of the digits chosen by the first spectator.

21. The Colliding Missiles

TWO MISSILES speed directly toward each other, one at 9,000 miles per hour and the other at 21,000 miles per hour. They start 1,317 miles apart. Without using pencil and paper, calculate how far apart they are one minute before they collide.

22. The Sliding Pennies

SIX PENNIES are arranged on a flat surface as shown in the top picture. The problem is to move them into the formation depicted at bottom in the smallest number of moves. Each move consists in sliding a penny, without disturbing any of the other pennies, to a new position in which it touches two others. The coins must remain flat on the surface at all times.

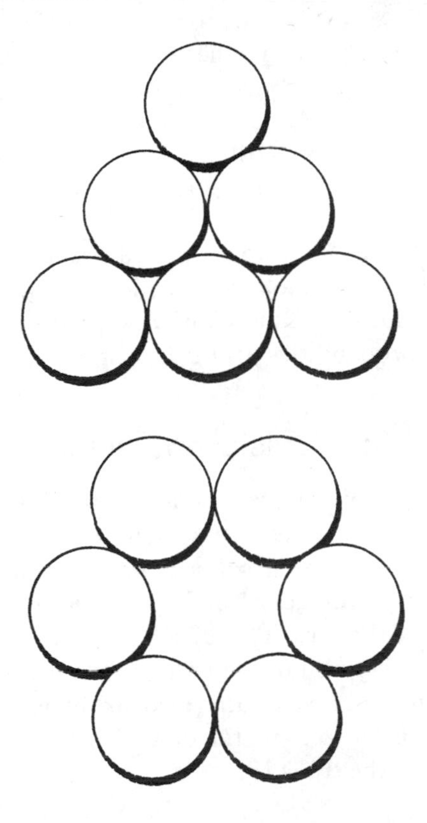

23. Handshakes and Networks

PROVE THAT at a recent convention of biophysicists the number of scientists in attendance who shook hands an odd number of times is even. The same problem can be expressed graphically as follows. Put as many dots (biophysicists) as you wish on a sheet of paper. Draw as many lines (handshakes) as you wish from any dot to any other dot. A dot can "shake hands" as often as you please, or not at all. Prove that the number of dots with an odd number of lines joining them is even.

24. The Triangular Duel

SMITH, BROWN and Jones agree to fight a pistol duel under the following unusual conditions. After drawing lots to determine who fires first, second and third, they take their places at the corners of an equilateral triangle. It is agreed that they will fire single shots in turn and continue in the same cyclic order until two of them are dead. At each turn the man who is firing may aim wherever he pleases. All three duelists know that Smith always hits his target, Brown is 80 per cent accurate and Jones is 50 per cent accurate.

Assuming that all three adopt the best strategy, and that no one is killed by a wild shot not intended for him, who has the best chance to survive? A more difficult question: What are the exact survival probabilities of the three men?

25. Crossing the Desert

AN UNLIMITED supply of gasoline is available at one edge of a desert 800 miles wide, but there is no source on the desert itself. A truck can carry enough gasoline to go 500 miles (this will be called one "load"), and it can build up its own refueling stations at any spot along the way. These caches may be any size, and it is assumed that there is no evaporation loss.

What is the minimum amount (in loads) of gasoline the truck will require in order to cross the desert? Is there a limit to the width of a desert the truck can cross?

26. Lord Dunsany's Chess Problem

ADMIRERS OF the Irish writer Lord Dunsany do not need to be told that he was fond of chess. (Surely his story "The Three Sailors' Gambit" is the funniest chess fantasy ever written.) Not generally known is the fact that he liked to invent bizarre chess problems which, like his fiction, combine humor and fantasy.

The problem depicted here was contributed by Dunsany to *The Week-End Problems Book*, compiled by Hubert Phillips. Its solution calls more for logical thought than skill at chess, although one does have to know the rules of the game. White is to play and mate in four moves. The position is one that could occur in actual play.

27. The Lonesome 8

THE MOST POPULAR problem ever published in *The American Mathematical Monthly*, its editors recently disclosed, is the following. It was contributed by P. L. Chessin of the Westinghouse Electric Corporation to the April 1954 issue.

"Our good friend and eminent numerologist, Professor Euclide Paracelso Bombasto Umbugio, has been busily engaged in testing on his desk calculator the 81×10^9 possible solutions to the problem of reconstructing the following exact long division in which the digits were indiscriminately replaced by x save in the quotient where they were almost entirely omitted:

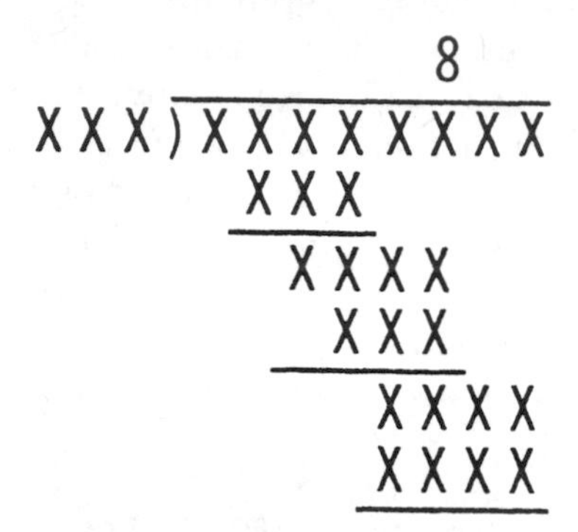

"Deflate the Professor! That is, reduce the possibilities to $(81 \times 10^9)^0$."

Because any number raised to the power of zero is one, the reader's task is to discover the unique reconstruction of the problem. The 8 is in correct position above the line, making it the third digit of a five digit-answer. The problem is easier than it looks, yielding readily to a few elementary insights.

28. Dividing the Cake

THERE IS a simple procedure by which two people can divide a cake so that each is satisfied he has at least half: One cuts and the other chooses. Devise a general procedure so that n persons can cut a cake into n portions in such a way that everyone is satisfied he has at least $1/n$ of the cake.

29. The Folded Sheet

MATHEMATICIANS have not yet succeeded in finding a formula for the number of different ways a road map can be folded, given n creases in the paper. Some notion of the complexity of this question can be gained from the following puzzle invented by the British puzzle expert Henry Ernest Dudeney.

Divide a rectangular sheet of paper into eight squares and number them on one side only, as shown in the top drawing. There are 40 different ways that this "map" can be folded along the ruled lines to form a square packet which has the "1" square face-up on top and all other squares beneath. The problem is to fold this sheet so that the squares are in serial order from 1 to 8, with the 1 face-up on top.

1	8	7	4
2	3	6	5

1	8	2	7
4	5	3	6

If you succeed in doing this, try the much more difficult task of doing the same thing with the sheet numbered in the manner pictured at the bottom of the illustration.

30. Water and Wine

A FAMILIAR chestnut concerns two beakers, one containing water, the other wine. A certain amount of water is transferred to the wine, then the same amount of the mixture is transferred back to the water. Is there now more water in the wine than there is wine in the water? The answer is that the two quantities are the same.

Raymond Smullyan writes to raise the further question: Assume that at the outset one beaker holds 10 ounces of water and the other holds 10 ounces of wine. By transferring three ounces back and forth any number of times, stirring after each transfer, is it possible to reach a point at which the percentage of wine in each mixture is the same?

31. The Absent-Minded Teller

AN ABSENT-MINDED bank teller switched the dollars and cents when he cashed a check for Mr. Brown, giving him dollars instead of cents, and cents instead of dollars. After buying a five-cent newspaper, Brown discovered that he had left exactly twice as much as his original check. What was the amount of the check?

32. Acute Dissection

GIVEN A TRIANGLE with one obtuse angle, is it possible to cut the triangle into smaller triangles, all of them acute? (An acute triangle is a triangle with three acute angles. A right angle is of course neither acute nor obtuse.) If this cannot be done, give a proof of impossibility. If it can be done, what is the smallest number of acute triangles into which any obtuse triangle can be dissected?

The illustration shows a typical attempt that leads nowhere. The triangle has been divided into three acute triangles, but the fourth is obtuse, so nothing has been gained by the preceding cuts.

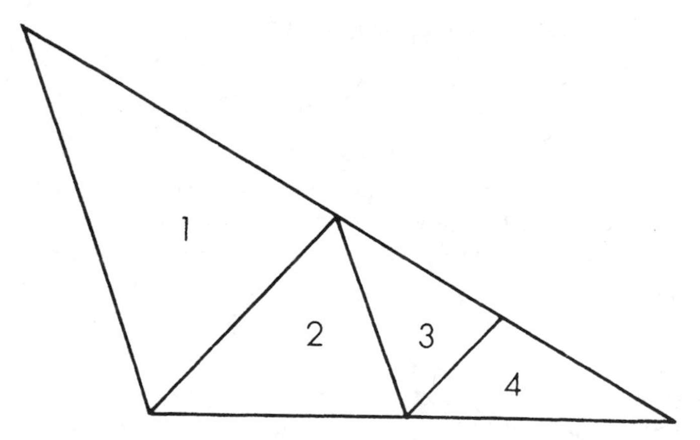

The problem (which came to me by way of Mel Stover of Winnipeg) is amusing because even the best mathematician is likely to be led astray by it and come to a wrong conclusion. My pleasure in working on it led me to ask myself a related question: "What is the smallest number of acute triangles into which a *square* can be dissected?" For days I was convinced that nine was the answer; then suddenly I saw how to reduce it to eight. I wonder how many readers can discover an eight-triangle solution, or perhaps an even better one. I am unable to prove that eight is the minimum, though I strongly suspect that it is.

33. How Long Is a "Lunar"?

In H. G. Wells's novel *The First Men in the Moon* our natural satellite is found to be inhabited by intelligent insect creatures who live in caverns below the surface. These creatures, let us assume, have a unit of distance that we shall call a "lunar." It was adopted because the moon's surface area, if expressed in square lunars, exactly equals the moon's volume in cubic lunars. The moon's diameter is 2,160 miles. How many miles long is a lunar?

34. The Game of Googol

In 1958 John H. Fox, Jr., of the Minneapolis-Honeywell Regulator Company, and L. Gerald Marnie, of the Massachusetts Institute of Technology, devised an unusual betting game which they call Googol. It is played as follows: Ask someone to take as many slips of paper as he pleases, and on each slip write a different positive number. The numbers may range from small fractions of 1 to a number the size of a "googol" (1 followed by a hundred 0's) or even larger. These slips are turned face down and shuffled over the top of a table. One at a time you turn the slips face up. The aim is to stop turning when you come to the number that you guess to be the largest of the series. You cannot go back and pick a previously turned slip. If you turn over all the slips, then of course you must pick the last one turned.

Most people will suppose the odds against your finding the highest number to be at least five to one. Actually if you adopt the best strategy, your chances are a little better than one in three. Two questions arise. First, what is the best strategy? (Note that this is not the same as asking for a strategy that will maximize the *value* of the selected number.) Second, if you follow this strategy, how can you calculate your chances of winning?

When there are only two slips, your chance of winning is obviously 1/2, regardless of which slip you pick. As the slips increase in number, the probability of winning (assuming that you use the best strategy) decreases, but the curve flattens quickly, and there is very little change beyond ten slips. The probability never drops below 1/3. Many players will suppose that they can make the task more difficult by choosing very large numbers, but a little reflection will show that the sizes of the numbers are irrelevant. It is only necessary that the slips bear numbers that can be arranged in increasing order.

The game has many interesting applications. For example, a girl decides to marry before the end of the year. She estimates that she will meet ten men who can be persuaded to propose, but once she has rejected a proposal, the man will not try again. What strategy should she follow to maximize her chances of accepting the top man of the ten, and what is the probability that she will succeed?

The strategy consists of rejecting a certain number of slips of paper (or proposals), then picking the next number that exceeds the highest number among the rejected slips. What is needed is a formula for determining how many slips to reject, depending on the total number of slips.

35. Marching Cadets and a Trotting Dog

A SQUARE FORMATION of Army cadets, 50 feet on the side, is marching forward at a constant pace. The company mascot, a small terrier, starts at the center of the rear rank [*position A in the illustration*], trots forward in a straight line to the center of the front rank [*position B*], then trots back again in a straight line to the center of the rear. At the instant he returns to position A, the cadets have advanced exactly 50 feet. Assuming that the dog trots at a constant speed and loses no time in turning, how many feet does he travel?

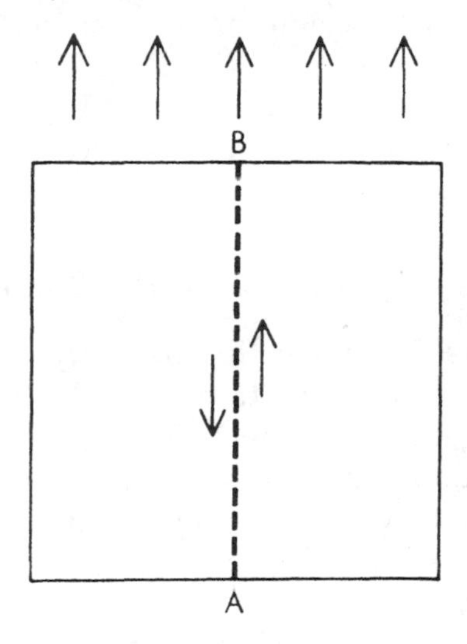

If you solve this problem, which calls for no more than a knowledge of elementary algebra, you may wish to tackle a much more difficult version proposed by the famous puzzlist Sam Loyd. (See *More Mathematical Puzzles of Sam Loyd,* Dover, 1960, page 103.) Instead of moving forward and back through the marching cadets, the mascot trots with constant speed around the *outside* of the square, keeping as close as possible to the square at all times. (For the problem we assume that he trots along the perimeter of the square.) As before, the formation has marched 50 feet by the time the dog returns to point A. How long is the dog's path?

36. White, Black and Brown

PROFESSOR MERLE WHITE of the mathematics department, Professor Leslie Black of philosophy, and Jean Brown, a young stenographer who worked in the university's office of admissions, were lunching together.

"Isn't it remarkable," observed the lady, "that our last names are Black, Brown and White and that one of us has black hair, one brown hair and one white."

"It is indeed," replied the person with black hair, "and have you noticed that not one of us has hair that matches his or her name?"

"By golly, you're right!" exclaimed Professor White.

If the lady's hair isn't brown, what is the color of Professor Black's hair?

37. The Plane in the Wind

AN AIRPLANE FLIES in a straight line from airport A to airport B, then back in a straight line from B to A. It travels with a constant engine speed and there is no wind. Will its travel time for the same round trip be greater, less or the same if, throughout both flights, at the same engine speed, a constant wind blows from A to B?

38. What Price Pets?

THE OWNER of a pet shop bought a certain number of hamsters and half that many pairs of parakeets. He paid $2 each for the hamsters and $1 for each parakeet. On every pet he placed a retail price that was an advance of 10 per cent over what he paid for it.

After all but seven of the creatures had been sold, the owner found that he had taken in for them an amount of money exactly equal to what he had originally paid for all of them. His potential profit, therefore, was represented by the combined retail value of the seven remaining animals. What was this value?

39. The Game of Hip

THE GAME of "Hip," so named because of the hipster's reputed disdain for "squares," is played on a six-by-six checkerboard as follows:

One player holds eighteen red counters; his opponent holds eighteen black counters. They take turns placing a single counter on any vacant cell of the board. Each tries to avoid placing his counters so that four of them mark the corners of a square. The square may be any size and tipped at any angle. There are 105 possible squares, four of which are shown in the illustration.

A player wins when his opponent becomes a "square" by forming one of the 105 squares. The game can be played on a board with actual counters, or with pencil and paper. Simply draw the board, then register moves by marking X's and O's on the cells.

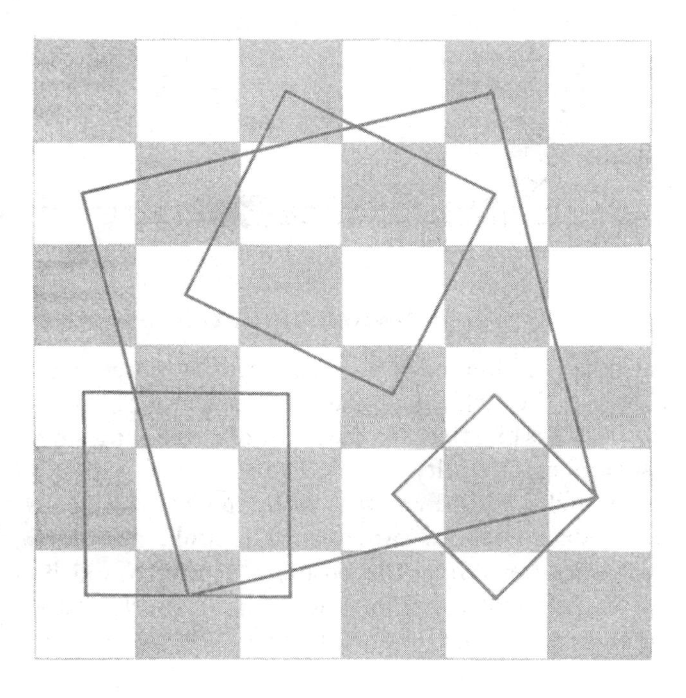

For months after I had devised this game I believed that it was impossible for a draw to occur in it. Then C. M. McLaury, a mathematics student at the University of Oklahoma, demonstrated that the game could end in a draw. The problem is to show how the game can be drawn by dividing the 36 cells into two sets of eighteen each so that no four cells of the same set mark the corners of a square.

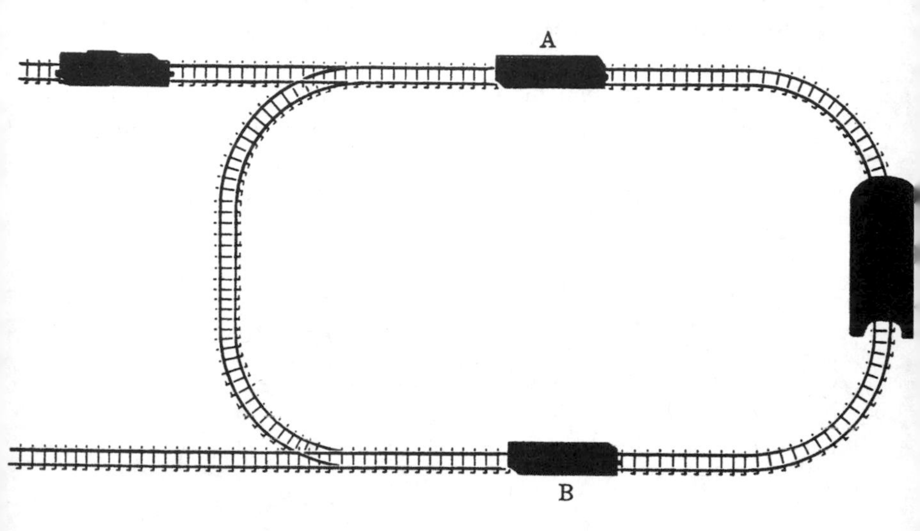

40. A Switching Puzzle

THE EFFICIENT switching of railroad cars often poses frustrating problems in the field of operations research. The switching puzzle shown is one that has the merit of combining simplicity with surprising difficulty.

The tunnel is wide enough to accommodate the locomotive but not wide enough for either car. The problem is to use the locomotive for switching the positions of cars A and B, then return the locomotive to its original spot. Each end of the locomotive can be used for pushing or pulling, and the two cars may, if desired, be coupled to each other.

The best solution is the one requiring the fewest operations. An "operation" is here defined as any movement of the locomotive between stops, assuming that it stops when it reverses direction, meets a car to push it or unhooks from a car it has been pulling. Movements of the two switches are not counted as operations.

A convenient way to work on the puzzle is to place a penny, a dime and a nickel on the illustration and slide them along the tracks, remembering that only the coin representing the locomotive can pass through the tunnel. In the illustration, the cars were drawn in positions too close to the switches. While working on the problem, assume that both cars are far enough east along the track so that there is ample space between each car and switch to accommodate both the locomotive and the other car.

No "flying switch" maneuvers are permitted. For example, you are not permitted to turn the switch quickly just after the engine has pushed an unattached car past it, so that the car goes one way and the engine, without stopping, goes another way.

41. Beer Signs on the Highway

SMITH DROVE at a steady clip along the highway, his wife beside him. "Have you noticed," he said, "that those annoying signs for Flatz beer seem to be regularly spaced along the road? I wonder how far apart they are."

Mrs. Smith glanced at her wrist watch, then counted the number of Flatz beer signs they passed in one minute.

"What an odd coincidence!" exclaimed Smith. "When you multiply that number by ten, it exactly equals the speed of our car in miles per hour."

Assuming that the car's speed is constant, that the signs are equally spaced and that Mrs. Smith's minute began and ended with the car midway between two signs, how far is it between one sign and the next?

42. The Sliced Cube and the Sliced Doughnut

AN ENGINEER, noted for his ability to visualize three-dimensional structure, was having coffee and doughnuts. Before he dropped a sugar cube into his cup, he placed the cube on the table and thought: If I pass a horizontal plane through the cube's center, the cross section will of course be a square. If I pass it vertically through the center and four corners of the cube, the cross section will be an oblong rectangle. Now suppose I cut the cube this way with the plane. . . . To his surprise, his mental image of the cross section was a regular hexagon.

How was the slice made? If the cube's side is half an inch, what is the side of the hexagon?

After dropping the cube into his coffee, the engineer turned his attention to a doughnut lying flat on a plate. "If I pass a plane horizontally through the center," he said to himself, "the cross section will be two concentric circles. If I pass the plane vertically through the center, the section will be two circles separated by the width of the hole. But if I turn the plane so. . . . " He whistled with astonishment. The section consisted of two perfect circles that intersected!

How was this slice made? If the doughnut is a perfect torus, three inches in outside diameter and with a hole one inch across, what are the diameters of the intersecting circles?

43. Bisecting Yin and Yang

Two mathematicians were dining at the Yin and Yang, a Chinese restaurant on West Third Street in Manhattan. They chatted about the symbol on the restaurant's menu, shown in the illustration.

"I suppose it's one of the world's oldest religious symbols," one of them said. "It would be hard to find a more attractive way to symbolize the great polarities of nature: good and evil, male and female, inflation and deflation, integration and differentiation."

"Isn't it also the symbol of the Northern Pacific Railway?"

"Yes. I understand that one of the chief engineers of the railroad saw the emblem on a Korean flag at the Chicago World's Fair in 1893 and urged his company to adopt it. He said it symbolized the extremes of fire and water that drove the steam engine."

"Do you suppose it inspired the construction of the modern baseball?"

"I wouldn't be surprised. By the way, did you know that there is an elegant method of drawing one straight line across the circle so that it exactly bisects the areas of the Yin and Yang?"

Assuming that the Yin and Yang are separated by two semicircles, show how each can be simultaneously bisected by the same straight line.

44. The Blue-Eyed Sisters

IF YOU HAPPEN to meet two of the Jones sisters (this assumes that the two are random selections from the set of all the Jones sisters), it is an exactly even-money bet that both girls will be blue-eyed. What is your best guess as to the total number of blue-eyed Jones sisters?

45. How Old Is the Rose-Red City?

TWO PROFESSORS, one of English and one of mathematics, were having drinks in the faculty club bar.

"It is curious," said the English professor, "how some poets can write one immortal line and nothing else of lasting value. John William Burgon, for example. His poems are so mediocre that no one reads them now, yet he wrote one of the most marvelous lines in English poetry: 'A rose-red city half as old as Time.' "

The mathematician, who liked to annoy his friends with improvised brainteasers, thought for a moment or two, then raised his glass and recited:

"A rose-red city half as old as Time.
One billion years ago the city's age
Was just two-fifths of what Time's age will be
A billion years from now. Can you compute
How old the crimson city is today?"

The English professor had long ago forgotten his algebra, so he quickly shifted the conversation to another topic, but readers of this book should have no difficulty with the problem.

46. Tricky Track

THREE HIGH SCHOOLS—Washington, Lincoln and Roosevelt—competed in a track meet. Each school entered one man, and one only, in each event. Susan, a student at Lincoln High, sat in the bleachers to cheer her boyfriend, the school's shot-put champion.

When Susan returned home later in the day, her father asked how her school had done.

"We won the shot-put all right," she said, "but Washington High won the track meet. They had a final score of 22. We finished with 9. So did Roosevelt High."

"How were the events scored?" her father asked.

"I don't remember exactly," Susan replied, "but there was a certain number of points for the winner of each event, a smaller number for second place and a still smaller number for third place. The numbers were the same for all events." (By "number" Susan of course meant a positive integer.)

"How many events were there altogether?"

"Gosh, I don't know, Dad. All I watched was the shot-put."

"Was there a high jump?" asked Susan's brother.

Susan nodded.

"Who won it?"

Susan didn't now.

Incredible as it may seem, this last question can be answered with only the information given. Which school won the high jump?

47. Termite and 27 Cubes

IMAGINE a large cube formed by gluing together 27 smaller wooden cubes of uniform size as shown. A termite starts at the center of the face of any one of the outside cubes and bores a path that takes him once through every cube. His movement is always parallel to a side of the large cube, never diagonal.

Is it possible for the termite to bore through each of the 26 outside cubes once and only once, then finish his trip by entering the central cube for the first time? If possible, show how it can be done; if impossible, prove it.

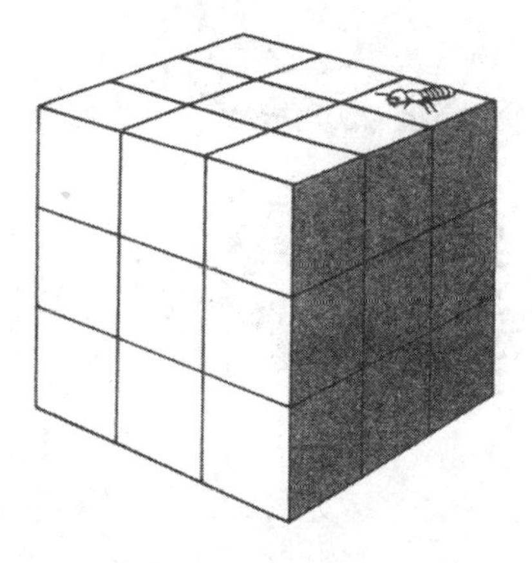

It is assumed that the termite, once it has bored into a small cube, follows a path entirely within the large cube. Otherwise, it could crawl out on the surface of the large cube and move along the surface to a new spot of entry. If this were permitted, there would, of course, be no problem.

48. Collating the Coins

ARRANGE THREE pennies and two dimes in a row, alternating the coins as shown. The problem is to change their positions to those shown at the bottom of the illustration in the shortest possible number of moves.

A move consists of placing the tips of the first and second fingers on any two touching coins, *one of which must be a penny and the other a dime,* then sliding the pair to another spot along the imaginary line shown in the illustration. The two coins in the pair must touch at all times. The coin at left in the pair must remain at left; the coin at right must remain at right. Gaps in the chain are allowed at the end of any move except the final one. After the last move the coins need not be at the same spot on the imaginary line that they occupied at the start.

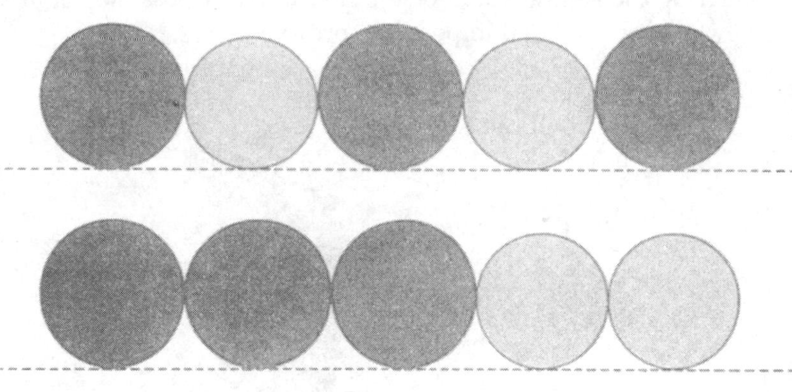

If it were permissible to shift two coins of the same kind, the puzzle could be solved easily in three moves: slide 1, 2 to left, fill the gap with 4, 5, then move 5, 3 from right to left end. But with the proviso that each shifted pair must include a dime and penny it is a baffling and pretty problem.

49. Time the Toast

EVEN THE SIMPLEST of household tasks can present complicated problems in operational research. Consider the preparation of three slices of hot buttered toast. The toaster is the old-fashioned type, with hinged doors on its two sides. It holds two pieces of bread at once but toasts each of them on one side only. To toast both sides it is necessary to open the doors and reverse the slices.

It takes three seconds to put a slice of bread into the toaster, three seconds to take it out and three seconds to reverse a slice without removing it. Both hands are required for each of these operations, which means that it is not possible to put in, take out or turn two slices simultaneously. Nor is it possible to butter a slice while another slice is being put into the toaster, turned or taken out. The toasting time for one side of a piece of bread is thirty seconds. It takes twelve seconds to butter a slice.

Each slice is buttered on one side only. No side may be buttered until it has been toasted. A slice toasted and buttered on one side may be returned to the toaster for toasting on its other side. The toaster is warmed up at the start. In how short a time can three slices of bread be toasted on both sides and buttered?

50. A Fixed-Point Theorem

ONE MORNING, exactly at sunrise, a Buddhist monk began to climb a tall mountain. The narrow path, no more than a foot or two wide, spiraled around the mountain to a glittering temple at the summit.

The monk ascended the path at varying rates of speed, stopping many times along the way to rest and to eat the dried fruit he carried with him. He reached the temple shortly before sunset. After several days of fasting and meditation he began his journey back along the same path, starting at sunrise and again walking at variable speeds with many pauses along the way. His average speed descending was, of course, greater than his average climbing speed.

Prove that there is a spot along the path that the monk will occupy on both trips at precisely the same time of day.

51. How Did Kant Set His Clock?

IT IS SAID that Immanuel Kant was a bachelor of such regular habits that the good people of Königsberg would adjust their clocks when they saw him stroll past certain landmarks.

One evening Kant was dismayed to discover that his clock had run down. Evidently his manservant, who had taken the day off, had forgotten to wind it. The great philosopher did not reset the hands because his watch was being repaired and he had no way of knowing the correct time. He walked to the home of his friend Schmidt, a merchant who lived a mile or so away, glancing at the clock in Schmidt's hallway as he entered the house.

After visiting Schmidt for several hours Kant left and walked home along the route by which he came. As always, he walked with a slow, steady gait that had not varied in twenty years. He had no notion of how long this return trip took. (Schmidt had recently moved into the area and Kant had not yet timed himself on this walk.) Nevertheless, when Kant entered his house, he immediately set his clock correctly.

How did Kant know the correct time?

52. Playing Twenty Questions when Probability Values Are Known

IN THE well-known game Twenty Questions one person thinks of an object, such as the Liberty Bell in Philadelphia or Lawrence Welk's left little toe, and another person tries to guess the object by asking no more than twenty questions, each answerable by yes or no. The best questions are usually those that divide the set of possible objects into two subsets as nearly equal in number as possible. Thus if a person has chosen as his "object" a number from 1 through 9, it can be guessed by this procedure in no more than four questions—possibly less. In twenty questions one can guess any number from 1 through 2^{20} (or 1,048,576).

Suppose that each of the possible objects can be given a different value to represent the probability that it has been chosen. For example, assume that a deck of cards consists of one ace of spades, two deuces of spades, three threes, and on up to nine nines, making 45 spade cards in all. The deck is shuffled; someone draws a card. You are to guess it by asking yes-no questions. How can you minimize the number of questions that you will probably have to ask?

53. Don't Mate in One

KARL FABEL, a German chess problemist, is responsible for this outrageous problem.

You are asked to find a move for white that will *not* result in an immediate checkmate of the black king.

54. Find the Hexahedrons

A POLYHEDRON is a solid bounded by plane polygons known as the faces of the solid. The simplest polyhedron is the tetrahedron, consisting of four faces, each a triangle [*left figure*]. A tetrahedron can have an endless variety of shapes, but if we regard its network of edges as a topological invariant (that is, we may alter the length of any edge and the angles at which edges meet but we must preserve the structure of the network), then there is only one basic type of tetrahedron. It is not possible, in other words, for a tetrahedron to have sides that are anything but triangles.

The five-sided polyhedron has two basic varieties [*middle and right figures*]. One is represented by the Great Pyramid of Egypt (four triangles on a quadrilateral base). The other is represented by a tetrahedron with one corner sliced off; three of its faces are quadrilaterals, two are triangles.

John McClellan, an artist in Woodstock, New York, asks this question: How many basic varieties of convex hexahedrons, or six-sided solids, are there altogether? (A solid is convex if each of its sides can be placed flat against a table top.) The cube is, of course, the most familiar example.

If you search for hexahedrons by chopping corners from simpler solids, you must be careful to avoid duplication. For example, the Great Pyramid, with its apex sliced off, has a skeleton that is topologically equivalent to that of the cube. Be careful also to avoid models that cannot exist without warped faces.

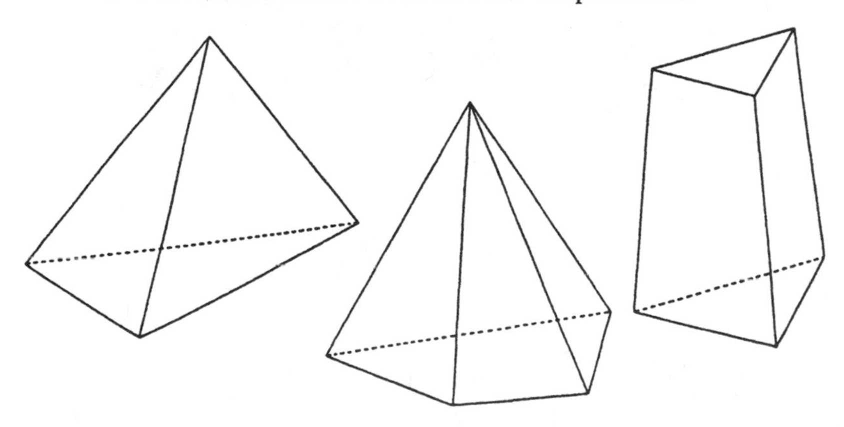

55. Out with the Onion

Arrange four paper matches on the table as shown in the left-hand figure. They represent a martini glass. A match head goes inside to indicate the onion of a Gibson cocktail. The puzzle is to move just *two* matches so that the glass is reformed, but the onion—which must stay where it is—winds up *outside* the glass. At the finish, the glass may be turned to the left or right, or even be upside down, but it must be exactly the same shape as before. The upper right figure is not a solution because the onion is still inside. The lower right figure doesn't work because *three* matches have been moved.

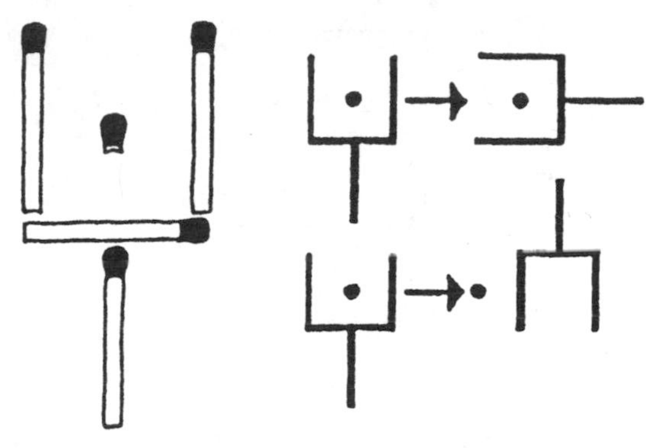

56. Cut Down the Cuts

You have taken up the hobby of jewelry making and you want to join the four pieces of silver chain shown at the left to form the circular bracelet shown at the right. Since it takes a bit of doing to cut open a link and weld it together again, you naturally want to cut as few links as possible. What is the minimum number of links you must cut to do the job?

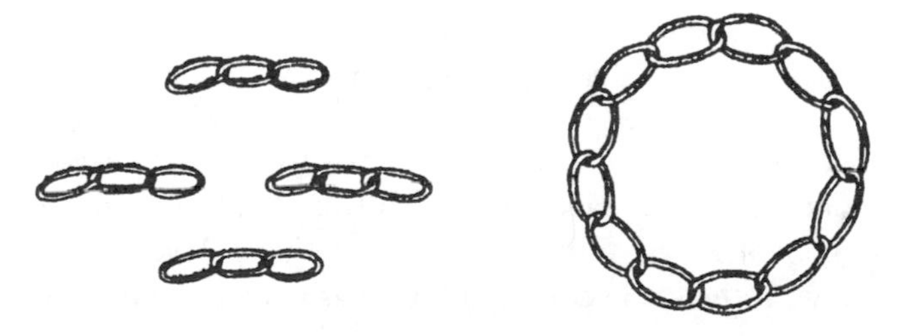

57. Dissection Dilemma

THE TWO TOP figures show how each of two shapes can be divided into four parts, all exactly alike. Your task is to divide the blank square into *five* parts, all identical in size and shape.

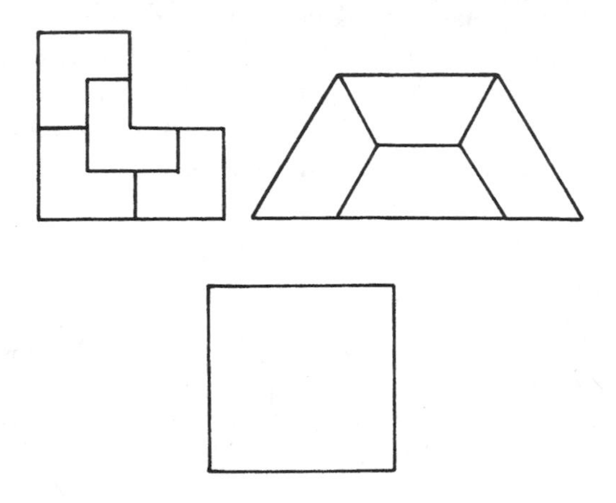

58. Interrupted Bridge

AFTER YOU have dealt about half the cards for a bridge game, the telephone rings. You put down the undealt cards to answer the phone. After you return, neither you nor anyone else can remember where the last card was dealt. No one has touched any of the dealt cards. Without counting the cards in any hand, or the number of cards yet to be dealt, how can you finish the deal rapidly and accurately, giving each player exactly the same cards he would have received if you hadn't been interrupted?

59. Dash It All!

SAUL AND SAL race each other for 100 yards. Sal wins by 10 yards. They decide to race again, but this time, to even things up, Sal begins 10 yards behind the start line. Assuming that both run with the same constant speed as before, who wins?

60. Move the Queen

PLACE a chess queen on a square next to a corner square of a chessboard as shown in the left-hand figure. The problem is to move the queen four times, making standard queen moves, so that she passes through all nine of the shaded squares. (A queen may move any number of squares in any direction: horizontally, vertically or diagonally.) The top figure shows one way to do it in six moves, but that's two too many.

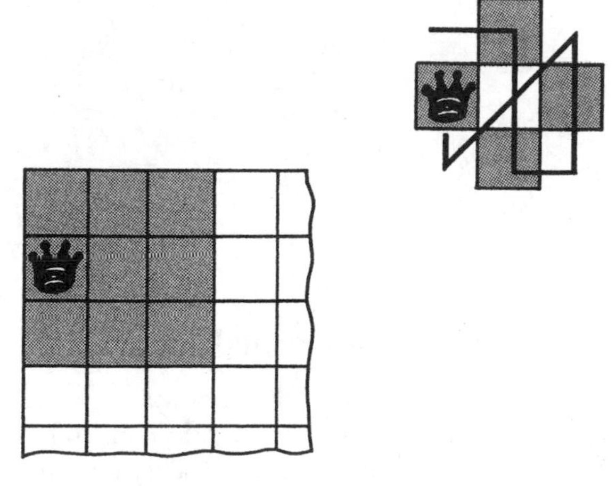

61. Read the Hieroglyphics

THE SEVEN symbols shown below look like some kind of ancient writing. But there is a meaning for each symbol, and if you can puzzle them out, you should have no trouble drawing in the square the next symbol of this curious sequence.

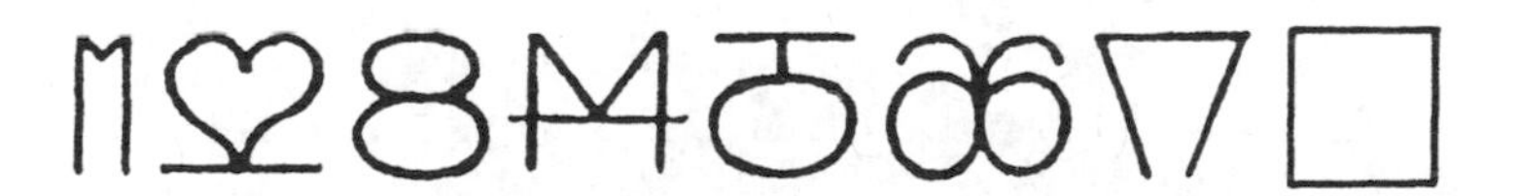

62. Crazy Cut

THIS ONE looks much easier than it is. You are to make one cut (or draw one line)—of course it needn't be straight—that will divide the figure into two identical parts.

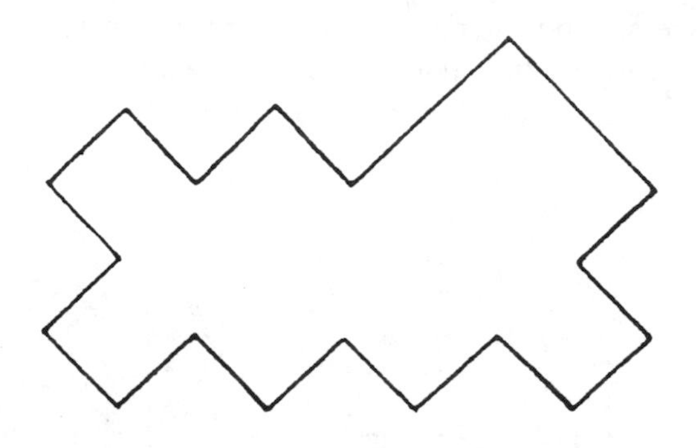

63. Find the Oddball

IF YOU HAVE two identical balls, one heavier than the other, you can easily determine which is heavier by putting them on opposite pans of a balance scale. If there are four balls, all the same weight except for one heavier one, you can find the heavier one in two weighings.

Suppose you have nine identical balls, one of which is heavier than the eight others. What is the smallest number of weighings needed for positively identifying the odd ball?

64. Big Cross-Out Swindle

CROSS OUT nine letters in such a way that the remaining letters spell a single word.

N A I S N I E N L G E L T E T W E O R R S D

65. Reverse the Dog

ARRANGE 13 paper matches to make a dog that faces to the left as in the diagram. By lowering the dog's tail to the top dotted line, then moving the bottom match of the dog's head to the other dotted line, you have changed the picture so that the dog is looking the opposite way. Unfortunately, this leaves the dog's tail (now on the left) slanting down instead of up.

Can you move just two matches to make the dog face to the right, but with his tail pointing upward as before?

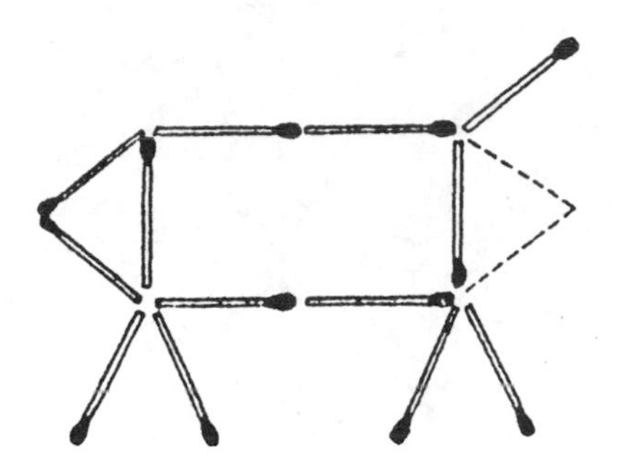

66. Funny Fold

SCOTT KIM, who tells me he invented this puzzle in sixth grade, has cut a large capital letter from a sheet of paper and given it a single fold.

It is easy to see that the letter could be an L, but that is *not* the letter Kim cut out. What letter is it?

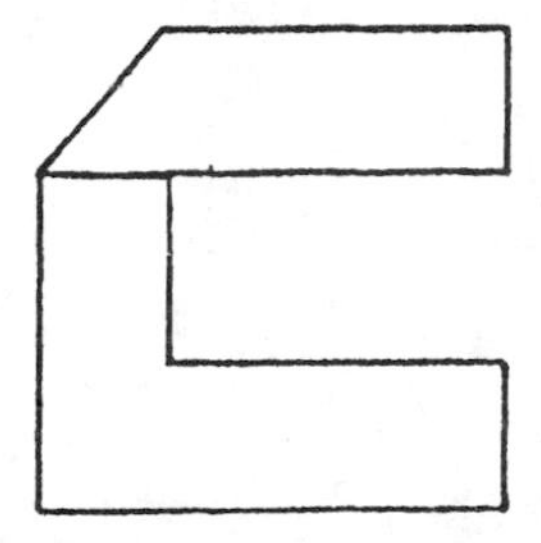

ANSWERS

1. Is there any other point on the globe, besides the North Pole, from which you could walk a mile south, a mile east, and a mile north and find yourself back at the starting point? Yes indeed; not just one point but an infinite number of them! You could start from any point on a circle drawn around the South Pole at a distance slightly more than $1 + 1/2\pi$ miles (about 1.16 miles) from the Pole—the distance is "slightly more" to take into account the curvature of the earth. After walking a mile south, your next walk of one mile east will take you on a complete circle around the Pole, and the walk one mile north from there will then return you to the starting point. Thus your starting point could be any one of the infinite number of points on the circle with a radius of about 1.16 miles from the South Pole. But this is not all. You could also start at points closer to the Pole, so that the walk east would carry you just twice around the Pole, or three times, and so on.

2. There are 88 winning first hands. They fall into two categories: (1) four tens and any other card (48 hands); (2) three tens and any of the following pairs from the suit not represented by a ten: A-9, K-9, Q-9, J-9, K-8, Q-8, J-8, Q-7, J-7, J-6 (40 hands).

3. It is impossible to cover the mutilated chessboard (with two opposite corner squares cut off) with 31 dominoes, and the proof is easy. The two diagonally opposite corners are of the same color. Therefore their removal leaves a board with two more squares of one color than of the other. Each domino covers two squares of opposite color, since only opposite colors are adjacent. After you have covered 60 squares with 30 dominoes, you are left with two uncovered squares of the same color. These two cannot be adjacent, therefore they cannot be covered by the last domino.

Suppose two cells of *opposite* color are removed from the chessboard. Can you prove that no matter which two cells are removed, the board can always be covered with 31 dominoes? For a simple, elegant proof that this is always possible, see my book *Aha! Insight* (W. H. Freeman, 1978, page 19).

4. If we require that the question be answerable by "yes" or "no," there are several solutions, all exploiting the same basic gimmick. For example, the logician points to one of the roads and says to the native, "If I were to ask you if this road leads to the village, would you say 'yes'?" The native is forced to give the right answer, even if he is a liar! If the road does lead to the village, the liar would say "no" to the direct question, but as the question is put, he lies and says he would respond "yes." Thus the logician can be certain that the road does lead to the village, whether the respondent is a truth-teller or a liar. On the other hand, if the road actually does not go to the village, the liar is forced in the same way to reply "no" to the inquirer's question.

A similar question would be, "If I asked a member of the other tribe whether this road leads to the village, would he say 'yes'?" To avoid the cloudiness that results from a question within a question, perhaps this phrasing (suggested by Warren C. Haggstrom, of Ann Arbor, Michigan) is best: "Of the two statements, 'You are a liar' and 'This road leads to the village,' is one and only one of them true?" Here again, a "yes" answer indicates it is the road, a "no" answer that it isn't, regardless of whether the native lies or tells the truth.

Dennis Sciama, Cambridge University cosmologist, and John McCarthy of Hanover, New Hampshire, called my attention to a delightful additional twist on the problem. "Suppose," Mr. McCarthy wrote (in a letter published in *Scientific American,* April 1957), "the logician knows that 'pish' and 'tush' are the native words for 'yes' and 'no' but has forgotten which is which, though otherwise he can speak the native language. He can still determine which road leads to the village.

"He points to one of the roads and asks, 'If I asked you whether the road I am pointing to is the road to the village would you say pish?' If the native replies, 'Pish,' the logician can conclude that the road pointed to is the road to the village even though he will still be in the dark as to whether the native is a liar or a truth-teller and as to whether 'pish' means yes or no. If the native says, 'Tush,' he may draw the opposite conclusion."

For hundreds of ingenious problems involving truth-tellers and liars, see the puzzle books by mathematician-logician Raymond M. Smullyan.

5. You can learn the contents of all three boxes by drawing just one marble. The key to the solution is your knowledge that the labels on all three of the boxes are incorrect. You must draw a marble from the box labeled "black-white." Assume that the marble drawn is black. You know then that the other marble in this box must be black also, otherwise the label would be correct. Since you have now identified the box containing two black marbles, you can at once tell the contents of the box marked "white-white": you know it cannot contain two white marbles, because its label has to be wrong; it cannot contain two black marbles, for you have identified that box; therefore it must contain one black and one white marble. The third box, of course, must then be the one holding two white marbles. You can solve the puzzle by the same reasoning if the marble you draw from the "black-white" box happens to be white instead of black.

6. There is no way to reduce the cuts to fewer than six. This is at once apparent when you focus on the fact that a cube has six sides. The saw cuts straight—one side at a time. To cut the one-inch cube at the center (the one which has no exposed surfaces to start with) must take six passes of the saw.

7. The answer to this puzzle is a simple matter of train schedules. While the Brooklyn and Bronx trains arrive equally often—at 10-minute intervals—it happens that their schedules are such that the Bronx train always comes to this platform one minute after the Brooklyn train. Thus the Bronx train will be the first to arrive only if the young man happens to come to the subway platform during this one-minute interval. If he enters the station at any other time—*i.e.*, during a nine-minute interval—the Brooklyn train will come first. Since the young man's arrival is random, the odds are nine to one for Brooklyn.

8. The commuter has walked for 55 minutes before his wife picks him up. Since they arrive home 10 minutes earlier than usual, this means that the wife has chopped 10 minutes from her usual travel time to and from the station, or five minutes from her travel time to the station. It follows that she met her husband five minutes before his usual pick-up time of five o'clock, or at 4:55. He started walking at four, therefore he walked for 55 minutes. The man's speed of walking, the wife's speed of driving and the distance between home and station are not needed for solving the problem. If you tried to solve it by juggling figures for these variables, you probably found the problem exasperating.

9. The counterfeit stack can be identified by a single weighing of coins. You take one coin from the first stack, two from the second, three from the third and so on to the entire 10 coins of the tenth stack. You then weigh the whole sample collection on the pointer scale. The excess weight of this collection, in number of grams, corresponds to the number of the counterfeit stack. For example, if the group of coins weighs seven grams more than it should, then the counterfeit stack must be the seventh one, from which you took seven coins (each weighing one gram more than a genuine half-dollar). Even if there had been an eleventh stack of ten coins, the procedure just described would still work, for no excess weight would indicate that the one remaining stack was counterfeit.

10. There are several different ways of placing the six cigarettes. The figure on the left shows the traditional solution as it is given in several old puzzle books.

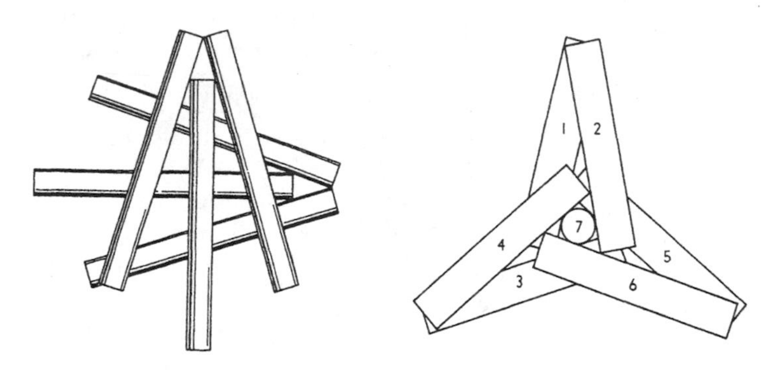

To my vast surprise, about fifteen readers discovered that *seven* cigarettes could also be placed so that each touched all of the others! This of course makes the older puzzle obsolete. The figure on the right, sent to me by George Rybicki and John Reynolds, graduate students in physics at Harvard, shows how it is done. "The diagram has been drawn," they write, "for the critical case where the ratio of length to diameter of the cigarettes is $\frac{7}{2}\sqrt{3}$. Here the points of contact occur right at the ends of the cigarettes. The solution obviously will work for any length-to-diameter ratio greater than $\frac{7}{2}\sqrt{3}$. Some observations on actual 'regular' size cigarettes give a ratio of about 8 to 1, which is, in fact, greater than $\frac{7}{2}\sqrt{3}$, so this is an acceptable solution." Note that if the center cigarette, pointing directly toward you in the diagram, is withdrawn, the remaining six provide a neat symmetrical solution of the original problem.

11. When the ferryboats meet for the first time [*top illustration*], the combined distance traveled by the boats is equal to the width of the river. When they reach the opposite shore, the combined distance is twice the width of the river; and when they meet the second time [*bottom figure*], the total distance is three times the river's width. Since the boats have been moving at a constant speed for the same period of time, it follows that each boat has gone three times as far as when they first met and had traveled a combined distance of one river-width. Since the white boat had traveled 720 yards when the first meeting occurred, its total distance at the time of the second meeting must be 3 × 720, or 2,160 yards. The bottom illustration shows clearly that this distance is 400 yards more than the river's width, so we subtract 400 from 2,160 to obtain 1,760 yards, or one mile, as the width of the river. The time the boats remained at their landings does not enter into the problem.

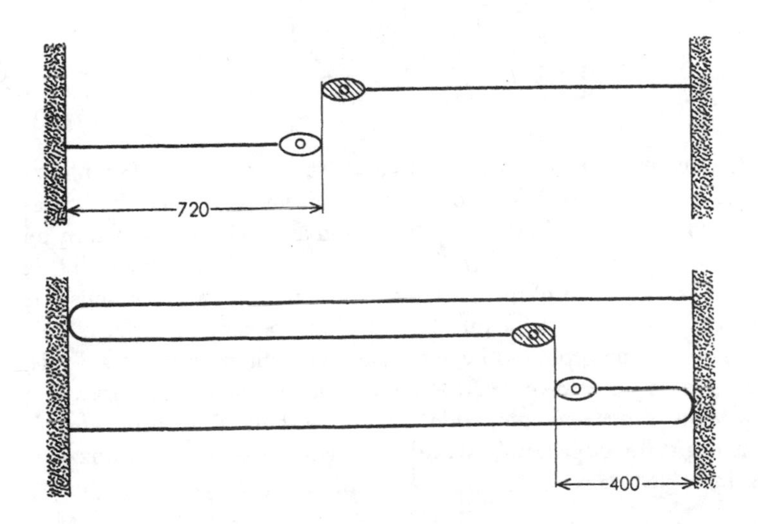

The problem can be approached in other ways. Many readers solved it as follows. Let x equal the river-width. On the first trip the ratio of distances traveled by the two boats is $x - 720:720$. On the second trip it is $2x - 400:x + 400$. These ratios are equal, so it is easy to solve for x.

12. Line AC is one diagonal of the rectangle. The other diagonal is clearly the 10-unit radius of the circle. Since the diagonals are equal, line AC is 10 units long.

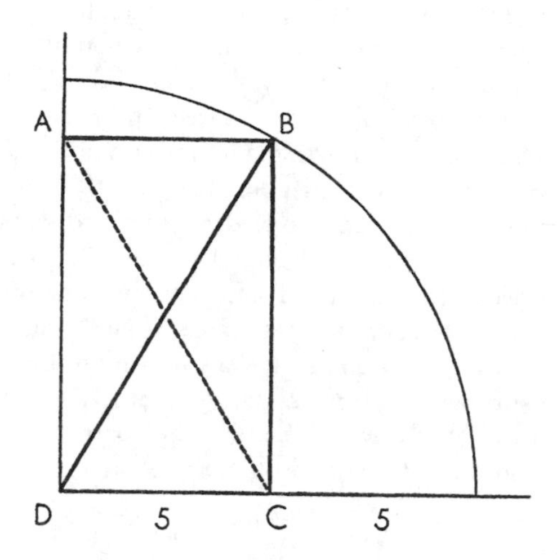

13. A continuous line that enters and leaves one of the rectangular spaces must of necessity cross two line segments. Since the spaces labeled A, B and C in the top illustration are each surrounded by an odd number of segments, it follows that an end of a line must be inside each if all segments of the network are crossed. But a continuous line has only two ends, so the puzzle is insoluble on a plane surface. This same reasoning applies if the network is on a sphere or on the side of a torus [*drawing at lower left*]. However, the network can be drawn on the torus [*drawing at lower right*] so that the hole of the torus is *inside* one of the three spaces, A, B and C. When this is done, the puzzle is easily solved.

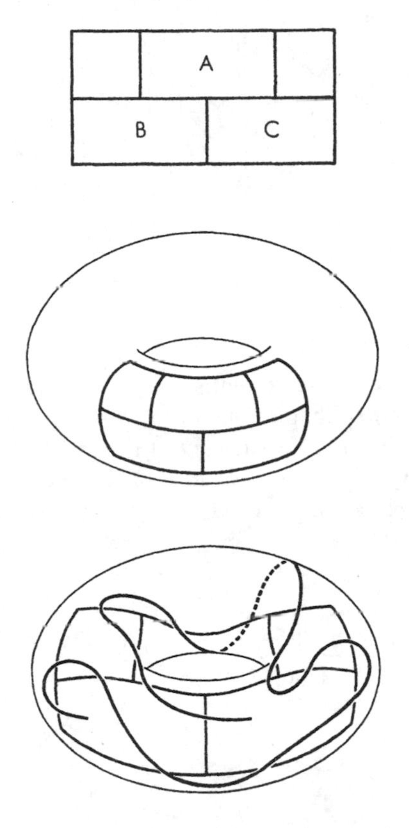

14. Twelve matches can be used to form a right triangle with sides of three, four and five units, as shown in the first illustration. This triangle will have an area of six square units. By altering the position of three matches as shown at right in the illustration, we remove two square units, leaving a polygon with an area of four.

This solution is the one to be found in many puzzle books. There are hundreds of other solutions. Elton M. Palmer, Oakmont, Pennsylvania, pointed out that each of the five tetrominoes (figures made with four squares) can provide the base for a large number of solutions. We simply add and subtract the same amount in triangular areas to accommodate all 12 matches. The second drawing depicts some representative samples, each row based on a different tetromino.

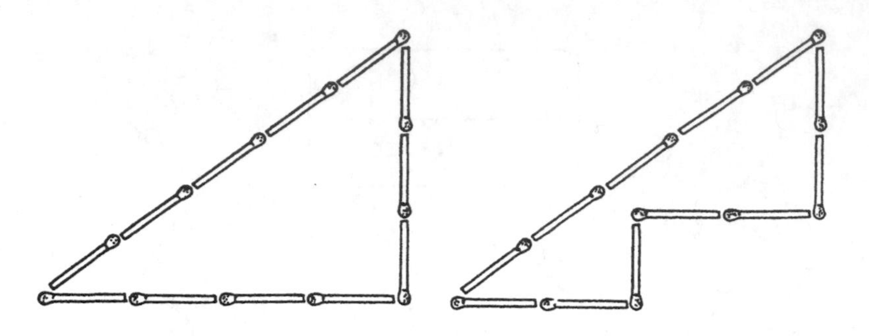

Eugene J. Putzer, staff scientist with the General Dynamics Corporation; Charles Shapiro, Oswego, New York; and Hugh J. Metz, Oak Ridge, Tennessee, suggested the star solution shown in the third drawing. By adjusting the width of the star's points you can produce any desired area between 0 and 11.196, the area of a regular dodecagon, the largest area possible with the 12 matches.

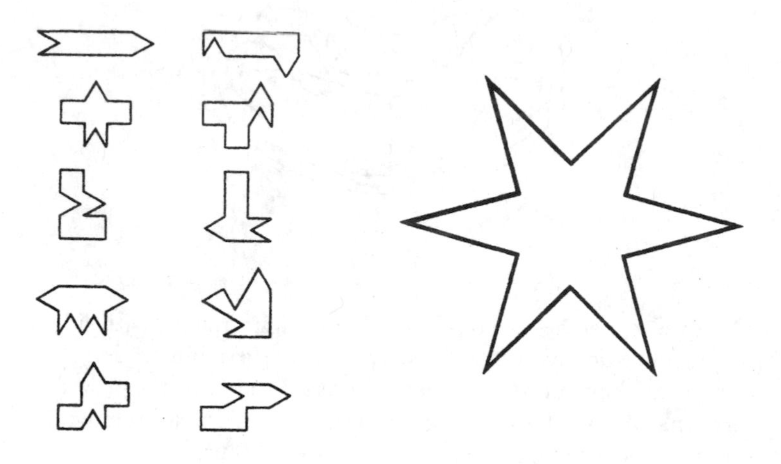

15. Without resorting to calculus, the problem can be solved as follows. Let R be the radius of the sphere. As the first illustration indicates, the radius of the cylindrical hole will then be the square root of $R^2 - 9$, and the altitude of the spherical caps at each end of the cylinder will be $R - 3$. To determine the residue after the cylinder and caps have been removed, we add the volume of the cylinder, $6\pi(R^2 - 9)$, to twice the volume of the spherical cap, and subtract the total from the volume of the sphere, $4\pi R^3/3$. The volume of the cap is obtained by the following formula, in which A stands for its altitude and r for its radius: $\pi A(3r^2 + A^2)/6$.

When this computation is made, all terms obligingly cancel out except 36π—the volume of the residue in cubic inches. In other words, the residue is constant regardless of the hole's diameter or the size of the sphere!

The earliest reference I have found for this beautiful problem is on page 86 of Samuel I. Jones's *Mathematical Nuts,* self-published, Nashville, 1932. A two-dimensional analog of the problem appears on page 93 of the same volume. Given the longest possible straight line that can be drawn on a circular track of any dimensions [*see second figure*], the area of the track will equal the area of a circle having the straight line as a diameter.

John W. Campbell, Jr., editor of *Astounding Science Fiction,* was one of several readers who solved the sphere problem quickly by reasoning adroitly as follows: The problem would not be given unless it has a unique solution. If it has a unique solution, the volume must be a constant which would hold even when the hole is reduced to zero radius. Therefore the residue must equal the volume of a sphere with a diameter of six inches, namely 36π.

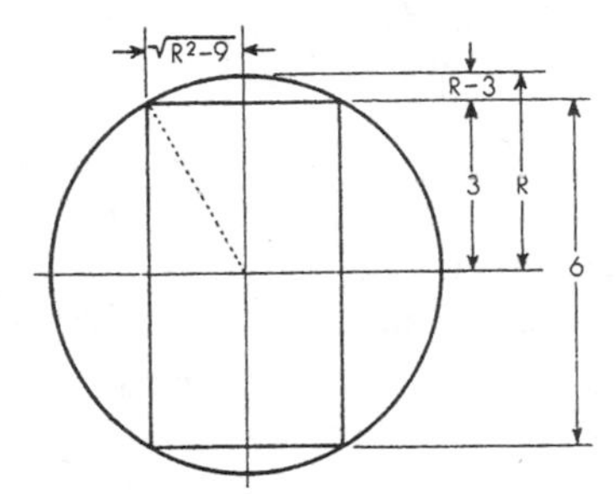

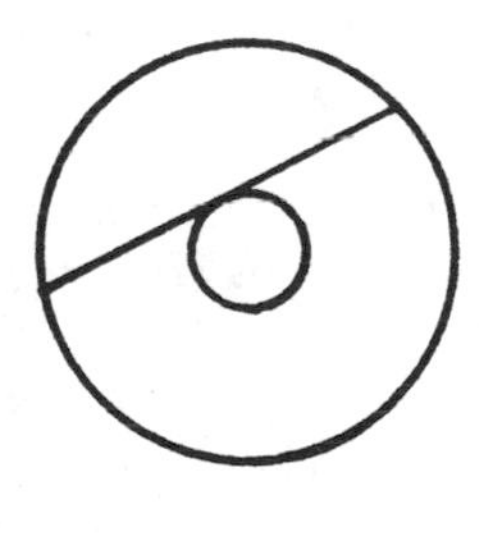

16. At any given instant the four bugs form the corners of a square which shrinks and rotates as the bugs move closer together. The path of each pursuer will therefore at all times be perpendicular to the path of the pursued. This tells us that as A, for example, approaches B, there is no component in B's motion which carries B toward or away from A. Consequently A will capture B in the same time that it would take if B had remained stationary. The length of each spiral path will be the same as the side of the square: 10 inches.

If three bugs start from the corners of an equilateral triangle, each bug's motion will have a component of 1/2 (the cosine of a 60-degree angle is 1/2) its velocity that will carry it toward its pursuer. Two bugs will therefore have a mutual approach speed of 3/2 velocity. The bugs meet at the center of the triangle after a time interval equal to twice the side of the triangle divided by three times the velocity, each tracing a path that is 2/3 the length of the triangle's side.

For a generalization of this problem to n bugs at the corners of n-sided polygons see Chapter 24 of my *Sixth Book of Mathematical Games from Scientific American* (W. H. Freeman, 1971).

17. When Jones began to work on the professor's problem he knew that each of the four families had a different number of children, and that the total number was less than 18. He further knew that the product of the four numbers gave the professor's house number. Therefore his obvious first step was to factor the house number into four different numbers which together would total less than 18. If there had been only one way to do this, he would have immediately solved the problem. Since he could not solve it without further information, we conclude that there must have been more than one way of factoring the house number.

Our next step is to write down all possible combinations of four different numbers which total less than 18, and obtain the products of each group. We find that there are many cases where more than one combination gives the same product. How do we decide which product is the house number?

The clue lies in the fact that Jones asked if there was more than one child in the smallest family. This question is meaningful only if the house number is 120, which can be factored as $1 \times 3 \times 5 \times 8$, $1 \times 4 \times 5 \times 6$, or $2 \times 3 \times 4 \times 5$. Had Smith answered "No," the problem would remain unsolved. Since Jones did solve it, we know the answer was "Yes." The families therefore contained 2, 3, 4 and 5 children.

This problem was originated by Lester R. Ford and published in the *American Mathematical Monthly*, March 1948, as Problem E776.

18. The heads of the twiddled bolts move neither inward nor outward. The situation is comparable to that of a person walking up an escalator at the same rate that it is moving down.

19. Three airplanes are quite sufficient to ensure the flight of one plane around the world. There are many ways this can be done, but the following seems to be the most efficient. It uses only five tanks of fuel, allows the pilots of two planes sufficient time for a cup of coffee and a sandwich before refueling at the base, and there is a pleasing symmetry in the procedure.

Planes A, B and C take off together. After going 1/8 of the distance, C transfers 1/4 tank to A and 1/4 to B. This leaves C with 1/4 tank, just enough to get back home.

Planes A and B continue another 1/8 of the way, then B transfers 1/4 tank to A. B now has 1/2 tank left, which is sufficient to get him back to the base where he arrives with an empty tank.

Plane A, with a full tank, continues until it runs out of fuel 1/4 of the way from the base. It is met by C which has been refueled at the base. C transfers 1/4 tank to A, and both planes head for home.

The two planes run out of fuel 1/8 of the way from the base, where they are met by refueled plane B. Plane B transfers 1/4 tank to each of the other two planes. The three planes now have just enough fuel to reach the base with empty tanks.

The entire procedure can be diagrammed as shown, where distance is the horizontal axis and time the vertical axis. The right and left edges of the chart should, of course, be regarded as joined.

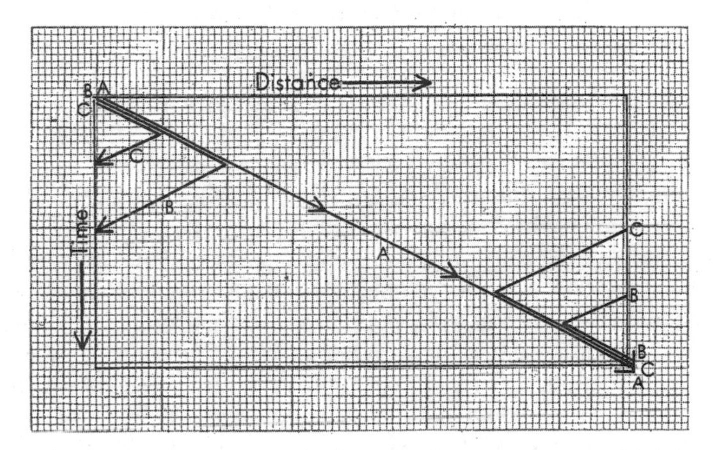

20. Writing a three-digit number twice is the same as multiplying it by 1,001. This number has the factors 7, 11 and 13, so writing the chosen number twice is equivalent to multiplying it by 7, 11 and 13. Naturally when the product is successively divided by these same three numbers, the final remainder will be the original number.

I presented this classical trick before hand and desk calculators became omnipresent. The trick is most effective when a single spectator uses a calculator to perform his divisions while your back is turned, only to see his originally chosen number mysteriously appear in the readout after the final division.

21. The two missiles approach each other with combined speeds of 30,000 miles per hour, or 500 miles per minute. By running the scene backward in time we see that one minute before the collision the missiles would have to be 500 miles apart.

22. Number the top coin in the pyramid 1, the coins in the next row 2 and 3, and those in the bottom row 4, 5 and 6. The following four moves are typical of many possible solutions: Move 1 to touch 2 and 4, move 4 to touch 5 and 6, move 5 to touch 1 and 2 below, move 1 to touch 4 and 5.

23. Because two people are involved in every handshake, the total score for everyone at the convention will be evenly divisible by two and therefore even. The total score for the men who shook hands an even number of times is, of course, also even. If we subtract this even score from the even total score of the convention, we get an even total score for those men who shook hands an odd number of times. Only an even number of odd numbers will total an even number, so we conclude that an even number of men shook hands an odd number of times.

There are other ways to prove the theorem; one of the best was sent to me by Gerald K. Schoenfeld, a medical officer in the U.S. Navy. At the start of the convention, before any handshakes have occurred, the number of persons who have shaken hands an odd number of times will be 0. The first handshake produces two "odd persons." From now on, handshakes are of three types: between two even persons, two odd persons, or one odd and one even person. Each even-even shake increases the number of odd persons by 2. Each odd-odd shake decreases the number of odd persons by 2. Each odd-even shake changes an odd person to even and an even person to odd, leaving the number of odd persons unchanged. There is no way, therefore, that the even number of odd persons can shift its parity; it must remain at all times an even number.

Both proofs apply to a graph of dots on which lines are drawn to connect pairs of dots. The lines form a network on which the number of dots that mark the meeting of an odd number of lines is even.

24. In the triangular pistol duel the poorest shot, Jones, has the best chance to survive. Smith, who never misses, has the second best chance. Because Jones's two opponents will aim at each other when their turns come, Jones's best strategy is to fire into the air until one opponent is dead. He will then get the first shot at the survivor, which gives him a strong advantage.

Smith's survival chances are the easiest to determine. There is a chance of 1/2 that he will get the first shot in his duel with Brown, in which case he kills him. There is a chance of 1/2 that Brown will shoot first, and since Brown is 4/5 accurate, Smith has a 1/5 chance of surviving. So Smith's chance of surviving Brown is 1/2 added to $1/2 \times 1/5 = 3/5$. Jones, who is accurate half the time, now gets a crack at Smith. If he misses, Smith kills him, so Smith has a survival chance of 1/2 against Jones. Smith's over-all chance of surviving is therefore $3/5 \times 1/2 = 3/10$.

Brown's case is more complicated because we run into an infinite series of possibilities. His chance of surviving against Smith is 2/5 (we saw above that Smith's survival chance against Brown was 3/5, and since one of the two men must survive, we subtract 3/5 from 1 to obtain Brown's chance of surviving against Smith). Brown now faces fire from Jones. There is a chance of 1/2 that Jones will miss, in which case Brown has a 4/5 chance of killing Jones. Up to this point, then, his chance of killing Jones is $1/2 \times 4/5 = 4/10$. But there is a 1/5 chance that Brown may miss, giving Jones another shot. Brown's chance of surviving is 1/2; then he has a 4/5 chance of killing Jones again, so his chance of surviving on the second round is $1/2 \times 1/5 \times 1/2 \times 4/5 = 4/100$.

If Brown misses again, his chance of killing Jones on the third round will be 4/1,000; if he misses once more, his chance on the fourth round will be 4/10,000, and so on. Brown's total survival chance against Jones is thus the sum of the infinite series:

$$\frac{4}{10} + \frac{4}{100} + \frac{4}{1{,}000} + \frac{4}{10{,}000} \cdots$$

This can be written as the repeating decimal .444444 . . ., which is the decimal expansion of 4/9.

As we saw earlier, Brown had a 2/5 chance of surviving Smith; now we see that he has a 4/9 chance to survive Jones. His over-all survival chance is therefore $2/5 \times 4/9 = 8/45$.

Jones's survival chance can be determined in similar fashion, but of course we can get it immediately by subtracting Smith's chance, 3/10, and Brown's chance, 8/45, from 1. This gives Jones an over-all survival chance of 47/90.

The entire duel can be conveniently graphed by using the tree diagram shown. It begins with only two branches because Jones passes if he has the first shot, leaving only two equal possibilities: Smith shooting first or Brown shooting first, with intent to kill. One branch of the tree goes on endlessly. The over-all survival chance of an individual is computed as follows:

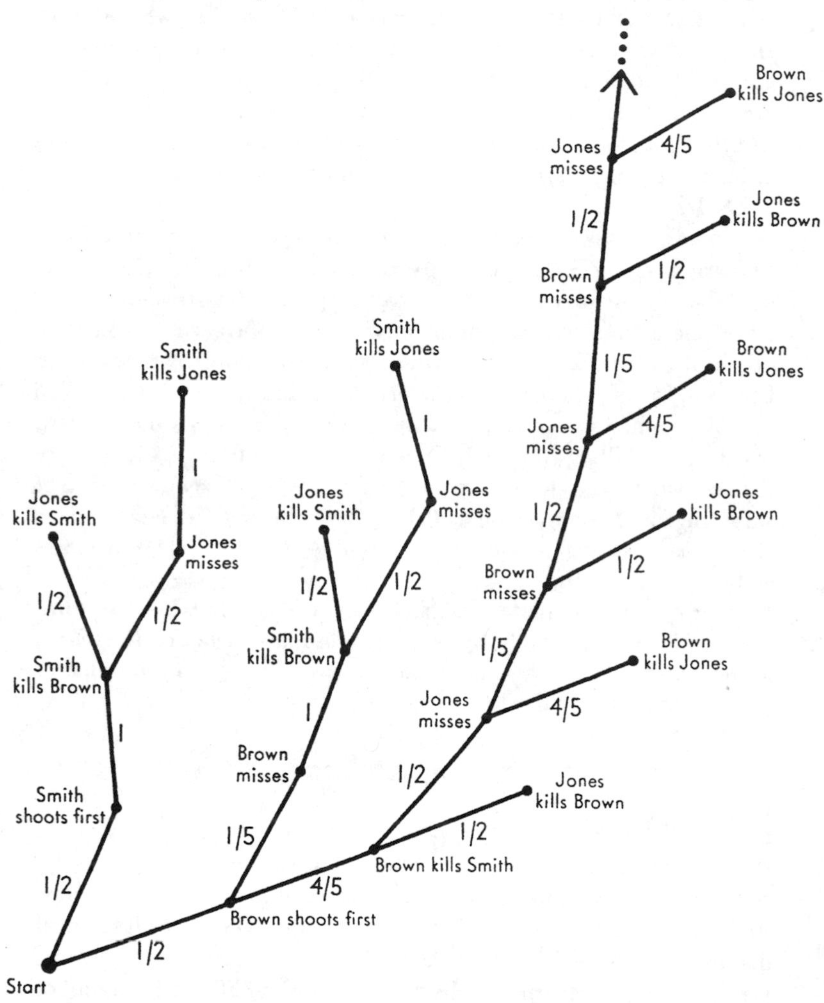

1. Mark all the ends of branches at which the person is sole survivor.

2. Trace each end back to the base of the tree, multiplying the probabilities of each segment as you go along. The product will be the probability of the event at the end of the branch.

3. Add together the probabilities of all the marked end-point events. The sum will be the over-all survival probability for that person.

In computing the survival chances of Brown and Jones, an infinite number of end-points are involved, but it is not difficult to see from the diagram how to formulate the infinite series that is involved in each case.

25. The following analysis of the desert-crossing problem appeared in an issue of *Eureka,* a publication of mathematics students at the University of Cambridge. Five hundred miles will be called a "unit"; gasoline sufficient to take the truck 500 miles will be called a "load"; and a "trip" is a journey of the truck in either direction from one stopping point to the next.

Two loads will carry the truck a maximum distance of 1 and 1/3 units. This is done in four trips by first setting up a cache at a spot 1/3 unit from the start. The truck begins with a full load, goes to the cache, leaves 1/3 load, returns, picks up another full load, arrives at the cache and picks up the cache's 1/3 load. It now has a full load, sufficient to take it the remaining distance to one unit.

Three loads will carry the truck 1 and 1/3 plus 1/5 units in a total of nine trips. The first cache is 1/5 unit from the start. Three trips put 6/5 loads in the cache. The truck returns, picks up the remaining full load and arrives at the first cache with 4/5 load in its tank. This, together with the fuel in the cache, makes two full loads, sufficient to carry the truck the remaining 1 and 1/3 units, as explained in the preceding paragraph.

We are asked for the minimum amount of fuel required to take the truck 800 miles. Three loads will take it 766 and 2/3 miles (1 and 1/3 plus 1/5 units), so we need a third cache at a distance of 33 and 1/3 miles (1/15 unit) from the start. In five trips the truck can build up this cache so that when the truck reaches the cache at the end of the seventh trip, the combined fuel of truck and cache will be three loads. As we have seen, this is sufficient to take the truck the remaining distance of 766 and 2/3 miles. Seven trips are made between starting point and first cache, using 7/15 load of gasoline. The three loads of fuel that remain are just sufficient for the rest of the way, so the total amount of gasoline consumed will be 3 and 7/15, or a little more than 3.46 loads. Sixteen trips are required.

Proceeding along similar lines, four loads will take the truck a distance of 1 and 1/3 plus 1/5 plus 1/7 units, with three caches located at the boundaries of these distances. The sum of this infinite series diverges as the number of loads increases; therefore the truck can cross a desert of any width. If the desert is 1,000 miles across, seven caches, 64 trips and 7.673 loads of gasoline are required.

Hundreds of letters were received on this problem, giving general solutions and interesting sidelights. Cecil G. Phipps, professor of mathematics at the University of Florida, summed matters up succinctly as follows:

"The general solution is given by the formula:

$$d = m\,(1 + 1/3 + 1/5 + 1/7 + \ldots),$$

where d is the distance to be traversed and m is the number of miles per load of gasoline. The number of depots to be established is one less than the number of terms in the series needed to exceed the value of d. One load of gasoline is used in the travel between each pair of stations. Since the series is divergent, any distance can be reached by this method although the amount of gasoline needed increases exponentially.

"If the truck is to return eventually to its home station, the formula becomes:

$$d = m\,(1/2 + 1/4 + 1/6 + 1/8 + \ldots)$$

This series is also divergent and the solution has properties similar to those for the one-way trip."

Many readers called attention to three previously published discussions of the problem:

"The Jeep Problem: A More General Solution." C. G. Phipps in the *American Mathematical Monthly,* Vol. 54, No. 8, pages 458–462, October 1947.

"Crossing the Desert." G. G. Alway in the *Mathematical Gazette,* Vol. 41, No. 337, page 209, October 1947.

Problem in Logistics: The Jeep Problem. Olaf Helmer. Project Rand Report No. RA-15015, December 1, 1946.

26. The key to Lord Dunsany's chess problem is the fact that the black queen is not on a black square as she must be at the start of a game. This means that the black king and queen have moved, and this could have happened only if some black pawns have moved. Pawns cannot move backward, so we are forced to conclude that the black pawns reached their present positions from the other side of the board! With this in mind, it is easy to discover that the white knight on the right has an easy mate in four moves.

White's first move is to jump his knight at the lower right corner of the board to the square just above his king. If black moves the upper left knight to the rook's file, white mates in two more moves. Black can, however, delay the mate one move by first moving his knight to the bishop's file instead of the rook's. White jumps his knight forward and right to the bishop's file, threatening mate on the next move. Black moves his knight forward to block the mate. White takes the knight with his queen, then mates with his knight on the fourth move.

27. In long division, when two digits are brought down instead of one, there must be a zero in the quotient. This occurs twice, so we know at once that the quotient is $x080x$. When the divisor is multiplied by the quotient's last digit, the product is a four-digit number. The quotient's last digit must therefore be 9, because eight times the divisor is a three-digit number.

The divisor must be less than 125 because eight times 125 is 1,000, a four-digit number. We now can deduce that the quotient's first digit must be more than 7, for seven times a divisor less than 125 would give a product that would leave more than two digits after it was subtracted from the first four digits in the dividend. This first digit cannot be 9 (which gives a four-digit number when the divisor is multiplied by it), so it must be 8, making the full quotient 80809.

The divisor must be more than 123 because 80809 times 123 is a seven-digit number and our dividend has eight digits. The only number between 123 and 125 is 124. We can now reconstruct the entire problem as follows:

```
          80809
    124 )10020316
          992
          ----
          1003
           992
          ----
           1116
           1116
           ----
```

28. Several procedures have been devised by which n persons can divide a cake in n pieces so that each is satisfied that he has at least $1/n$ of the cake. The following system has the merit of leaving no excess bits of cake.

Suppose there are five persons: A, B, C, D, E. A cuts off what he regards as 1/5 of the cake and what he is content to keep as his share. B now has the privilege, if he thinks A's slice is more than 1/5, of reducing it to what he thinks is 1/5 by cutting off a portion. Of course if he thinks it is 1/5 or less, he does not touch it. C, D and E in turn now have the same privilege. The last person to touch the slice keeps it as his share. Anyone who thinks that this person got less than 1/5 is naturally pleased because it means, in his eyes, that more than 4/5 remains. The remainder of the cake, including any cut-off pieces, is now divided among the remaining four persons in the same manner, then among three. The final division is made by one person cutting and the other choosing. The procedure is clearly applicable to any number of persons.

For a discussion of this and other solutions, see the section "Games of Fair Division," pages 363–368, in *Games and Decisions,* by R. Duncan Luce and Howard Raiffa, John Wiley and Sons, Inc., 1957.

The problem of dividing a cake between *n* persons so that each person is convinced he has his fair share has been the topic of many papers. Here are three:

"How to Cut a Cake Fairly," by L. E. Dubins and E. H. Spanier, in *American Mathematical Monthly,* Vol. 68, January 1961, pages 1–17.

"Preferred Shares," by Dominic Olivastro, in *The Sciences,* March/April 1992, pages 52–54.

"An Envy-Free Cake Division Algorithm," by Steven J. Brams and Alan D. Taylor, preprint. More than fifty references are cited in the bibliography.

29. The first sheet is folded as follows. Hold it face down so that when you look down on it the numbered squares are in this position:

$$\frac{2365}{1874}$$

Fold the right half on the left so that 5 goes on 2, 6 on 3, 4 on 1 and 7 on 8. Fold the bottom half up so that 4 goes on 5 and 7 on 6. Now tuck 4 and 5 between 6 and 3, and fold 1 and 2 under the packet.

The second sheet is first folded in half the long way, the numbers outside, and held so that 4536 is uppermost. Fold 4 on 5. The right end of the strip (squares 6 and 7) is pushed between 1 and 4, then bent around the folded edge of 4 so that 6 and 7 go between 8 and 5, and 3 and 2 go between 1 and 4.

For more puzzles involving paper folding see "The Combinatorics of Paper Folding," in my *Wheels, Life and Other Mathematical Amusements* (W. H. Freeman, 1983).

30. Regardless of how much wine is in one beaker and how much water is in the other, and regardless of how much liquid is transferred back and forth at each step (provided it is not all of the liquid in one beaker), it is impossible to reach a point at which the percentage of wine in each mixture is the same. This can be shown by a simple inductive argument. If beaker A contains a higher concentration of wine than beaker B, then a transfer from A to B will leave A with the higher concentration. Similarly a transfer from B to A—from a weaker to a stronger mixture—is sure to leave B weaker. Since every transfer is one of these two cases, it follows that beaker A must always contain a mixture with a higher percentage of wine than B. The only way to equalize the concentrations is by pouring all of one beaker into the other.

31. To determine the value of Brown's check, let x stand for the dollars and y for the cents. The problem can now be expressed by the following equation: $100y + x - 5 = 2(100x + y)$. This reduces to $98y - 199x = 5$, a Diophantine equation with an infinite number of integral solutions. A solution by the standard method of continued fractions gives as the lowest values in positive integers: $x = 31$ and $y = 63$, making Brown's check \$31.63. This is a unique answer to the problem because the next lowest values are: $x = 129$, $y = 262$, which fails to meet the requirement that y be less than 100.

There is a much simpler approach to the problem and many readers wrote to tell me about it. As before, let x stand for the dollars on the check, y for the cents. After buying his newspaper, Brown has left $2x + 2y$. The change that he has left, from the x cents given him by the cashier, will be $x - 5$.

We know that y is less than 100, but we don't know yet whether it is less than 50 cents. If it is less than 50 cents, we can write the following equations:

$$2x = y$$
$$2y = x - 5$$

If y is 50 cents or more, then Brown will be left with an amount of cents ($2y$) that is a dollar or more. We therefore have to modify the above equations by taking 100 from $2y$ and adding 1 to $2x$. The equations become:

$$2x + 1 = y$$
$$2y - 100 = x - 5$$

Each set of simultaneous equations is easily solved. The first set gives x a minus value, which is ruled out. The second set gives the correct values.

32. A number of readers sent "proofs" that an obtuse triangle cannot be dissected into acute triangles, but of course it can. The illustration shows a seven-piece pattern that applies to any obtuse triangle.

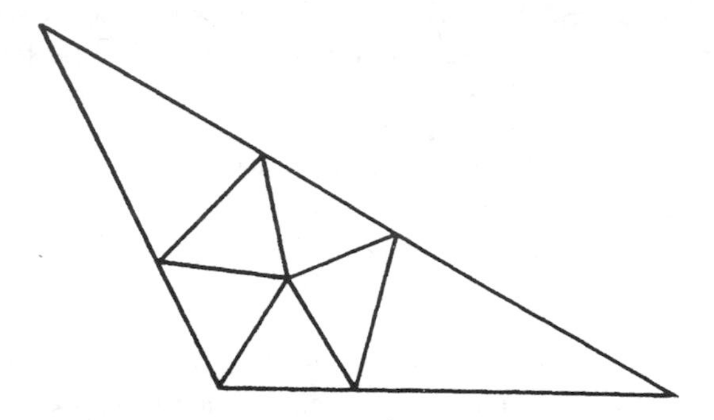

It is easy to see that seven is minimal. The obtuse angle must be divided by a line. This line cannot go all the way to the other side, for then it would form another obtuse triangle, which in turn would have to be dissected, consequently the pattern for the large triangle would not be minimal. The line dividing the obtuse angle must, therefore, terminate at a point *inside* the triangle. At this vertex, at least five lines must meet, otherwise the angles at this vertex would not all be acute. This creates the inner pentagon of five triangles, making a total of seven triangles in all. Wallace Manheimer, a Brooklyn high school teacher at the time, gave this proof as his solution to problem E1406 in *American Mathematical Monthly,* November 1960, page 923. He also showed how to construct the pattern for any obtuse triangle.

The question arises: Can any obtuse triangle be dissected into seven acute *isosceles* triangles? The answer is no. Verner E. Hoggatt, Jr., and Russ Denman (*American Mathematical Monthly,* November 1961, pages 912–913) proved that eight such triangles are sufficient for all obtuse triangles, and Free Jamison (*ibid.,* June–July 1962, pages 550–552) proved that eight are also necessary. These articles can be consulted for details as to conditions under which less than eight-piece patterns are possible. A right triangle and an acute nonisosceles triangle can each be cut into nine acute isosceles triangles, and an acute isosceles triangle can be cut into four congruent acute isosceles triangles similar to the original.

A square, I discovered, can be cut into eight acute triangles as shown. If the dissection has bilateral symmetry, points P and P′ must lie within the shaded area determined by the four semicircles. Donald L. Vanderpool pointed out in a letter that asymmetric distortions of the pattern are possible with point P anywhere outside the shaded area provided it remains outside the two large semicircles.

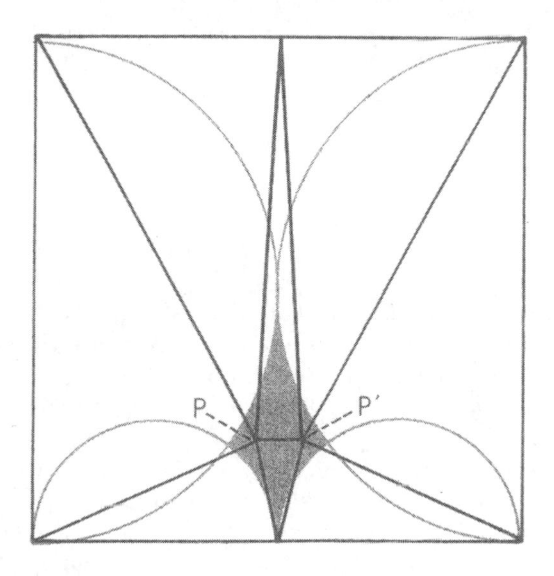

About 25 readers sent proofs, with varying degrees of formality, that the eight-piece dissection is minimal. One, by Harry Lindgren, appeared in *Australian Mathematics Teacher,* Vol. 18, pages 14–15, 1962. His proof also shows that the pattern, aside from the shifting of points P and P′ as noted above, is unique.

H. S. M. Coxeter pointed out the surprising fact that for any rectangle, even though its sides differ in length by an arbitrarily small amount, the line segment PP′ can be shifted to the center to give the pattern both horizontal and vertical symmetry.

Free Jamison found in 1968 that a square can be divided into ten acute isosceles triangles. See *The Fibonacci Quarterly* (December 1968) for a proof that a square can be dissected into any number of acute isosceles triangles equal or greater than 10.

33. The volume of a sphere is $4\pi/3$ times the cube of the radius. Its surface is 4π times the square of the radius. If we express the moon's radius in "lunars" and assume that its surface in square lunars equals its volume in cubic lunars, we can determine the length of the radius simply by equating the two formulas and solving for the value of the radius. Pi cancels out on both sides, and we find that the radius is three lunars. The moon's radius is 1,080 miles, so a lunar must be 360 miles.

34. Regardless of the number of slips involved in the game of Googol, the probability of picking the slip with the largest number (assuming that the best strategy is used) never drops below .367879. This is the reciprocal of e, and the limit of the probability of winning as the number of slips approaches infinity.

If there are ten slips (a convenient number to use in playing the game), the probability of picking the top number is .398. The strategy is to turn three slips, note the largest number among them, then pick the next slip that exceeds this number. In the long run you stand to win about two out of every five games.

What follows is a compressed account of a complete analysis of the game by Leo Moser and J. R. Pounder of the University of Alberta. Let n be the number of slips and p the number rejected before picking a number larger than any on the p slips. Number the slips serially from 1 to n. Let $k + 1$ be the number of the slip bearing the largest number. The top number will not be chosen unless k is equal to or greater than p (otherwise it will be rejected among the first p slips), and then only if the highest number from 1 to k is also the highest number from 1 to p (otherwise *this* number will be chosen before the top number is reached). The probability of finding the top number in case it is on the $k + 1$ slip is p/k, and the probability that the top number actually is on the $k + 1$ slip is $1/n$. Since the largest number can be on only one slip, we can write the following formula for the probability of finding it:

$$\frac{p}{n}\left(\frac{1}{p} + \frac{1}{p+1} + \frac{1}{p+2} \cdots\cdots + \frac{1}{n-1}\right)$$

Given a value for n (the number of slips) we can determine p (the number to reject) by picking a value for p that gives the greatest value to the above expression. As n approaches infinity, p/n approaches $1/e$, so a good estimate of p is simply the nearest integer to n/e. The general strategy, therefore, when the game is played with n slips, is to let n/e numbers go by, then pick the next number larger than the largest number on the n/e slips passed up.

This assumes, of course, that a player has no knowledge of the range of the numbers on the slips and therefore no basis for knowing whether a single number is high or low within the range. If one *has* such knowledge, the analysis does not apply. For example, if the game is played with the numbers on ten one-dollar bills, and your first draw is a bill with a number that begins with 9, your best strategy is to keep the bill. For similar reasons, the strategy in playing Googol is not strictly applicable to the unmarried girl problem, as many readers pointed out, because the girl presumably has a fair knowledge of the range in value of her suitors, and has certain standards in mind. If the first man who proposes comes very close to her ideal, wrote Joseph P. Robinson, "she would have rocks in her head if she did not accept at once."

Fox and Marnie apparently hit independently on a problem that had occurred to others a few years before. A number of readers said they had heard the problem before 1958—one recalled working on it in 1955—but I was unable to find any published reference to it. The problem of maximizing the *value* of the selected object (rather than the chance of getting the object of highest value) seems first to have been proposed by the famous mathematician Arthur Cayley in 1875. (See Leo Moser, "On a Problem of Cayley," in *Scripta Mathematica,* September–December 1956, pages 289–292.)

35. Let 1 be the length of the square of cadets and also the time it takes them to march this length. Their speed will also be 1. Let x be the total distance traveled by the dog and also its speed. On the dog's forward trip his speed relative to the cadets will be $x - 1$. On the return trip his speed relative to the cadets will be $x + 1$. Each trip is a distance of 1 (relative to the cadets), and the two trips are completed in unit time, so the following equation can be written:

$$\frac{1}{x-1} + \frac{1}{x+1} = 1$$

This can be expressed as the quadratic: $x^2 - 2x - 1 = 0$, for which x has the positive value of $1 + \sqrt{2}$. Multiply this by 50 to get the final answer: 120.7+ feet. In other words, the dog travels a total distance equal to the length of the square of cadets plus that same length times the square root of 2.

Loyd's version of the problem, in which the dog trots *around* the moving square, can be approached in exactly the same way. I paraphrase a clear, brief solution sent by Robert F. Jackson of the Computing Center at the University of Delaware.

As before, let 1 be the side of the square and also the time it takes the cadets to go 50 feet. Their speed will then also be 1. Let x be the distance traveled by the dog and also his speed. The dog's speed with respect to the speed of the square will be $x - 1$ on his forward trip, $\sqrt{x^2 - 1}$ on each of his two transverse trips, and $x + 1$ on his backward trip. The circuit is completed in unit time, so we can write this equation:

$$\frac{1}{x-1} + \frac{2}{\sqrt{x^2-1}} + \frac{1}{x+1} = 1$$

This can be expressed as the quartic equation: $x^4 - 4x^3 - 2x^2 + 4x + 5 = 0$. Only one positive real root is not extraneous: 4.18112+. We multiply this by 50 to get the desired answer: 209.056+ feet.

Theodore W. Gibson, of the University of Virginia, found that the first form of the above equation can be written as follows, simply by taking the square root of each side:

$$\frac{1}{\sqrt{x-1}} + \frac{1}{\sqrt{x+1}} = 1$$

which is remarkably similar to the equation for the first version of the problem.

Many readers sent analyses of variations of this problem: a square formation marching in a direction parallel to the square's diagonal, formations of regular polygons with more than four sides, circular formations, rotating formations, and so on. Thomas J. Meehan and David Salsburg each pointed out that the problem is the same as that of a destroyer making a square search pattern around a moving ship, and showed how easily it could be solved by vector diagrams on what the Navy calls a "maneuvering board."

36. The assumption that the "lady" is Jean Brown, the stenographer, quickly leads to a contradiction. Her opening remark brings forth a reply from the person with black hair, therefore Brown's hair cannot be black. It also cannot be brown, for then it would match her name. Therefore it must be white. This leaves brown for the color of Professor Black's hair and black for Professor White. But a statement by the person with black hair prompts an exclamation from White, so they cannot be the same person.

It is necessary to assume, therefore, that Jean Brown is a man. Professor White's hair can't be white (for then it would match his or her name), nor can it be black because he (or she) replies to the black-haired person. Therefore it must be brown. If the lady's hair isn't brown, then Professor White is not a lady. Brown is a man, so Professor Black must be the lady. Her hair can't be black or brown, so she must be a platinum blonde.

37. Since the wind boosts the plane's speed from A to B and retards it from B to A, one is tempted to suppose that these forces balance each other so that total travel time for the combined flights will remain the same. This is not the case, because the time during which the plane's speed is boosted is shorter than the time during which it is retarded, so the over-all effect is one of retardation. The total travel time in a wind of constant speed and direction, regardless of the speed or direction, is always greater than if there were no wind.

38. Let x be the number of hamsters originally purchased and also the number of parakeets. Let y be the number of hamsters among the seven unsold pets. The number of parakeets among the seven will then be $7 - y$. The number of hamsters sold (at a price of \$2.20 each, which is a markup of 10 per cent over cost) will be $x - y$, and the number of parakeets sold (at \$1.10 each) will be $x - 7 + y$.

The cost of the pets is therefore $2x$ dollars for the hamsters and x dollars for the parakeets—a total of $3x$ dollars. The hamsters that were sold brought $2.2\ (x - y)$ dollars and the parakeets sold brought $1.1\ (x - 7 + y)$ dollars—a total of $3.3x - 1.1y - 7.7$ dollars.

We are told that these two totals are equal, so we equate them and simplify to obtain the following Diophantine equation with two integral unknowns:

$$3x = 11y + 77$$

Since x and y are positive integers and y is not more than 7, it is a simple matter to try each of the eight possible values (including zero) for y to determine which of them makes x also integral. There are only two such values: 5 and 2. Each would lead to a solution of the problem were it not for the fact that the parakeets were bought in pairs. This eliminates 2 as a value for y because it would give x (the number of parakeets purchased) the odd value of 33. We conclude therefore that y is 5.

A complete picture can now be drawn. The shop owner bought 44 hamsters and 22 pairs of parakeets, paying altogether \$132 for them. He sold 39 hamsters and 21 pairs of parakeets for a total of \$132. There remained five hamsters worth \$11 retail and two parakeets worth \$2.20 retail—a combined value of \$13.20, which is the answer to the problem.

39. The illustration shows the finish of a drawn game of Hip. This beautiful, hard-to-find solution was first discovered by C. M. McLaury, a mathematics student at the University of Oklahoma to whom I had communicated the problem by way of Richard Andree, one of his professors.

Two readers (William R. Jordan, Scotia, New York, and Donald L. Vanderpool, Towanda, Pennsylvania) were able to show, by an exhaustive enumeration of possibilities, that the solution is unique except for slight variations in the four border cells indicated by arrows. Each cell may be either color, provided all four are not the same color, but since each player is limited in the game to eighteen pieces, two of these cells must be one color, two the other color. They are arranged here so that no matter how the square is turned, the pattern is the same when inverted.

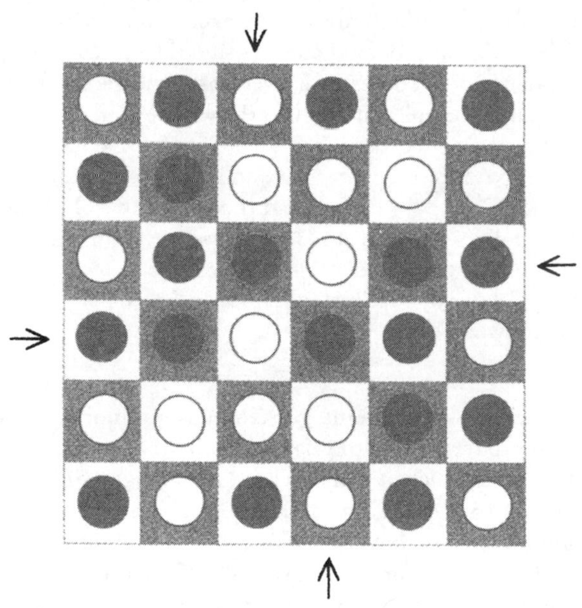

The order-6 board is the largest on which a draw is possible. This was proved in 1960 by Robert I. Jewett, then a graduate student at the University of Oregon. He was able to show that a draw is impossible on the order-7, and since all higher squares contain a seven-by-seven subsquare, draws are clearly impossible on them also.

As a playable game, Hip on an order-6 board is strictly for the squares. David H. Templeton, professor of chemistry at the University of California's Lawrence Radiation Laboratory in Berkeley, pointed out that the second player can always force a draw by playing a simple symmetry strategy. He can either make each move so that it matches his opponent's last move by reflection across a parallel bisector of the board, or by a 90-degree rotation about the board's center. (The latter strategy could lead to the draw depicted.) An alternate strategy is to play in the corresponding opposite cell on a line from the opponent's last move and across the center of the board. Second-player draw strategies were also sent by Allan W. Dickinson, Richmond Heights, Missouri, and Michael Merritt, a student at Texas A. & M. College. These strategies apply to all even-order fields, and since no draws are possible on such fields higher than 6, the strategy guarantees a win for the second player on all even-order boards of 8 or higher. Even on the order-6, a reflection strategy across a parallel bisector is sure to win, because the unique draw pattern does not have that type of symmetry.

Symmetry play fails on odd-order fields because of the central cell. Since nothing is known about strategies on odd-order boards, the order-7 is the best field for actual play. It cannot end in a draw, and no one at present knows whether the first or second player wins if both sides play rationally.

In 1963 Walter W. Massie, a civil engineering student at Worcester Polytechnic Institute, devised a Hip-playing program for the IBM 1620 digital computer, and wrote a term paper about it. The program allows the computer to play first or second on any square field of orders 4 through 10. The computer takes a random cell if it moves first. On other plays, it follows a reflection strategy except when a reflected move forms a square, then it makes random choices until it finds a safe cell.

On all square fields of order n, the number of different squares that can be formed by four cells is $(n^4 - n^2)/12$. The derivation of this formula, as well as a formula for rectangular boards, is given in Harry Langman, *Play Mathematics,* Hafner, 1962, pages 36–37.

As far as I know, no studies have been made of comparable "triangle-free" colorings on triangular lattice fields.

40. The locomotive can switch the positions of cars A and B, and return to its former spot, in 16 operations:

1. Locomotive moves right, hooks to car A.
2. Pulls A to bottom.
3. Pushes A to left, unhooks.

4. Moves right.
5. Makes a clockwise circle through tunnel.
6. Pushes B to left. All three are hooked.
7. Pulls A and B to right.
8. Pushes A and B to top. A is unhooked from B.
9. Pulls B to bottom.
10. Pushes B to left, unhooks.
11. Circles counterclockwise through tunnel.
12. Pushes A to bottom.
13. Moves left, hooks to B.
14. Pulls B to right.
15. Pushes B to top, unhooks.
16. Moves left to original position.

This procedure will do the job even when the locomotive is not permitted to pull with its front end, provided that at the start the locomotive is placed with its back toward the cars.

Howard Grossman, New York City, and Moises V. Gonzalez, Miami, Florida, each pointed out that if the lower siding is eliminated completely, the problem can still be solved, although two additional moves are required, making 18 in all. Can the reader discover how it is done?

Many different train-switching puzzles can be found in the puzzle books by Sam Loyd, Ernest Dudeney and others. Two recent articles deal with this type of puzzle: A. K. Dewdney's Computer Recreations column in *Scientific American* (June 1987) and "Reversing Trains: A Turn of the Century Sorting Problem," by Nancy Amato *et al.*, *Journal of Algorithms,* Vol. 10, 1989, pages 413–428.

41. The curious thing about the problem of the Flatz beer signs is that it is not necessary to know the car's speed to determine the spacing of the signs. Let x be the number of signs passed in one minute. In an hour the car will pass $60x$ signs. The speed of the car, we are told, is $10x$ miles per hour. In $10x$ miles it will pass $60x$ signs, so in one mile it will pass $60x/10x$, or 6, signs. The signs therefore are 1/6 mile, or 880 feet, apart.

42. A cube, cut in half by a plane that passes through the midpoints of six sides as shown, produces a cross section that is a regular hexagon. If the cube is half an inch on the side, the side of the hexagon is $\sqrt{2}/4$ inch.

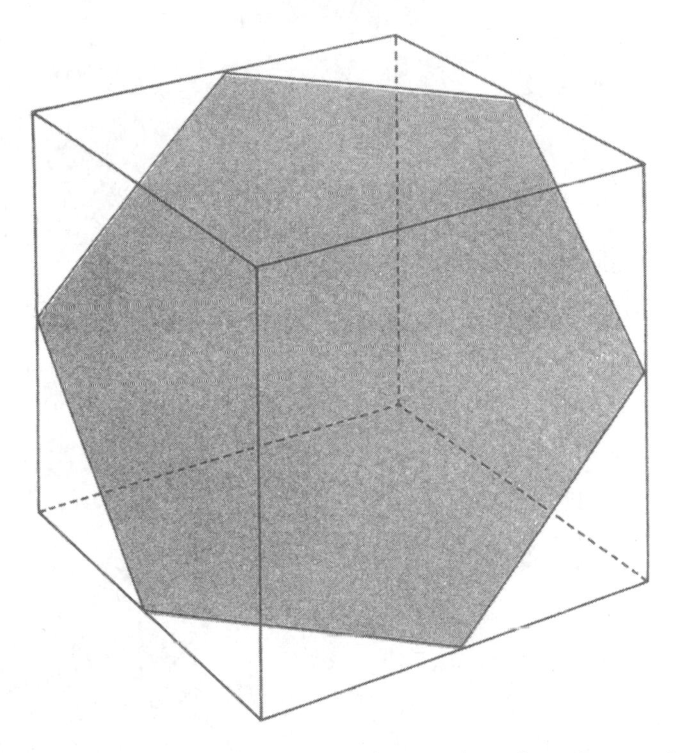

To cut a torus so that the cross section consists of two intersecting circles, the plane must pass through the center and be tangent to the torus above and below, as shown in the second picture (next page). If the torus and hole have diameters of three inches and one inch, each circle of the section will clearly have a diameter of two inches.

This way of slicing, and the two ways described earlier, are the only ways to slice a doughnut so that the cross sections are circular. Everett A. Emerson, in the electronics division of National Cash Register, Hawthorne, California, sent a full algebraic proof that there is no fourth way.

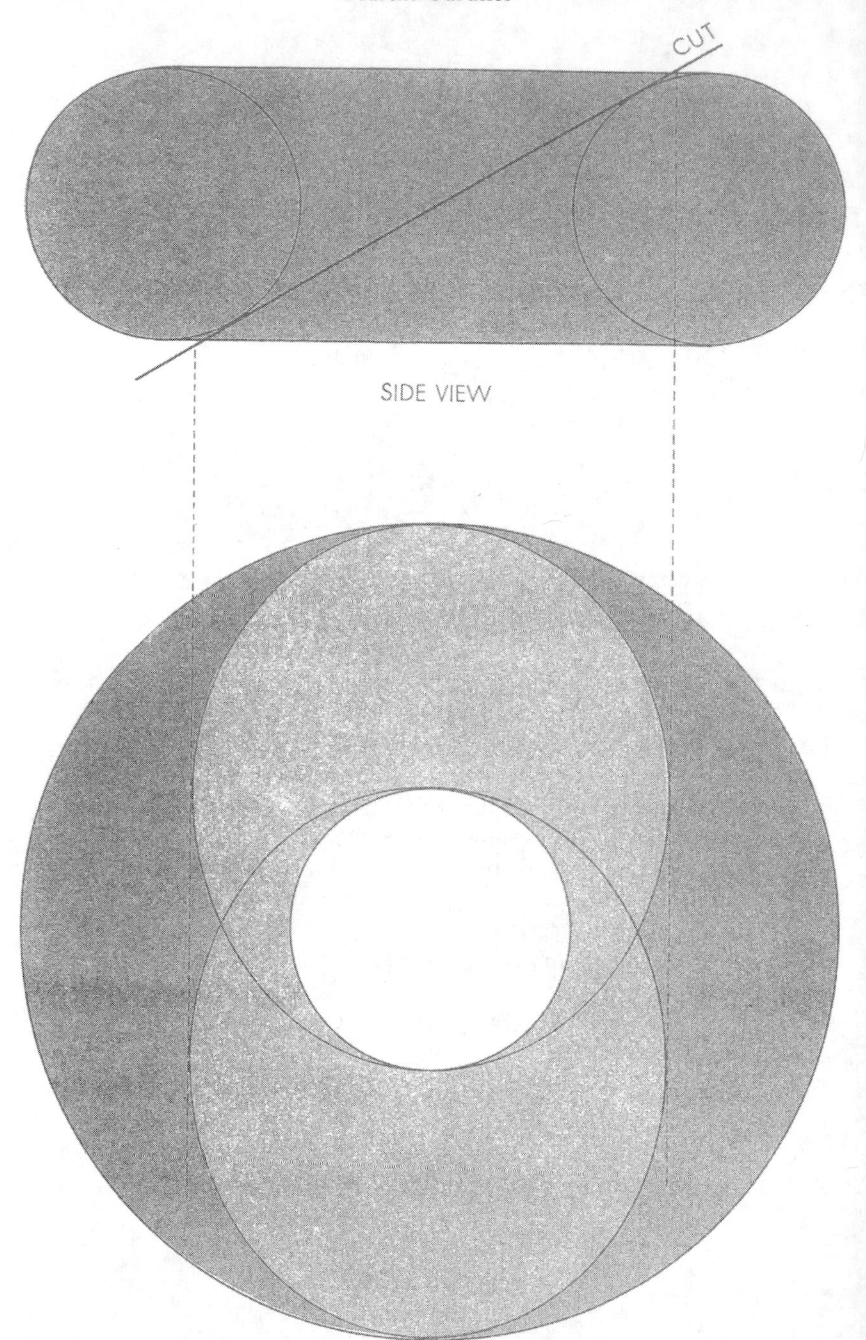
CUT
SIDE VIEW
TOP VIEW

43. The illustration shows how to construct a straight line that bisects both the Yin and the Yang. A simple proof is obtained by drawing the two broken semicircles. Circle K's diameter is half that of the monad; therefore its area is one-fourth that of the monad. Take region G from this circle, add H, and the resulting region is also one-fourth the monad's area. It follows that area G equals area H, and of course half of G must equal half of H. The bisecting line takes half of G away from circle K, but restores the same area (half of H) to the circle, so the black area below the bisecting line must have the same area as circle K. The small circle's area is one-fourth the large circle's area, therefore the Yin is bisected. The same argument applies to the Yang.

The foregoing proof was given by Henry Dudeney in his answer to problem 158, *Amusements in Mathematics* (Thomas Nelson & Sons, 1917; Dover reprint, 1958). After it appeared in *Scientific American,* four readers (A. E. Decae, F. J. Hooven, Charles W. Trigg and B. H. K. Willoughby) sent the following alternative proof, which is much simpler. Draw a horizontal diameter of the small circle K. The semicircle below this line has an area that is clearly 1/8 that of the large circle. Above the diameter is a 45-degree sector of the large circle (bounded by the small circle's horizontal diameter and the diagonal line) which also is obviously 1/8 the area of the large circle. Together, the semicircle and sector have an area of 1/4 that of the large circle, therefore the diagonal line must bisect both Yin and Yang. For ways of bisecting the Yin and Yang with curved lines, the reader is referred to Dudeney's problem, cited above, and Trigg's article, "Bisection of Yin and of Yang," in *Mathematics Magazine,* Vol. 34, No. 2, November–December 1960, pages 107–108.

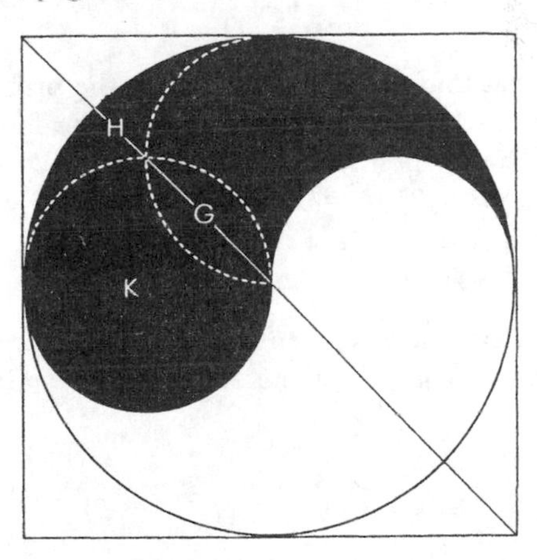

The Yin-Yang symbol (called the *T'ai-chi-t'u* in China and the *Tomoye* in Japan) is usually drawn with a small spot of Yin inside the Yang and a small spot of Yang inside the Yin. This symbolizes the fact that the great dualities of life are seldom pure; each usually contains a bit of the other. There is an extensive Oriental literature on the symbol. Sam Loyd, who bases several puzzles on the figure (*Sam Loyd's Cyclopedia of Puzzles,* page 26), calls it the Great Monad. The term "monad" is repeated by Dudeney, and also used by Olin D. Wheeler in a booklet entitled *Wonderland,* published in 1901 by the Northern Pacific Railway. Wheeler's first chapter is devoted to a history of the trademark, and is filled with curious information and color reproductions from Oriental sources.

For more on the symbol, see Schuyler Cammann, "The Magic Square of Three in Old Chinese Philosophy and Religion," *History of Religions,* Vol. 1, No. 1, Summer 1961, pages 37–80, my *New Ambidextrous Universe* (W. H. Freeman, 1979), pages 219–220, and George Sarton, *A History of Science,* Vol. 1 (Harvard University Press, 1952; Dover reprint, under the title *Ancient Science through the Golden Age of Greece,* 1993), page 11. Carl Gustav Jung cites some English references on the symbol in his introduction to the book of *I Ching* (1929), and there is a book called *The Chinese Monad: Its History and Meaning,* by Wilhelm von Hohenzollern, the date and publisher of which I do not know.

44. There are probably three blue-eyed Jones sisters and four sisters altogether. If there are n girls, of which b are blue-eyed, the probability that two chosen at random are blue-eyed is:

$$\frac{b(b-1)}{n(n-1)}$$

We are told that this probability is 1/2, so the problem is one of finding integral values for b and n that will give the above expression a value of 1/2. The smallest such values are $n = 4$, $b = 3$. The next highest values are $n = 21$, $b = 15$, but it is extremely unlikely that there would be as many as 21 sisters, so four sisters, three of them blue-eyed, is the best guess.

45. The rose-red city's age is seven billion years. Let x be the city's present age; y, the present age of Time. A billion years ago the city would have been $x - 1$ billion years old and a billion years from now Time's age will be $y + 1$. The data in the problem permit two simple equations:

$$2x = y$$

$$x - 1 = \frac{2}{5}(y + 1)$$

These equations give x, the city's present age, a value of seven billion years; and y, Time's present age, a value of fourteen billion years. The problem presupposes a "Big Bang" theory of the creation of the cosmos.

46. There is space only to suggest the procedure by which it can be shown that Washington High won the high jump event in the track meet involving three schools. Three different positive integers provide points for first, second and third place in each event. The integer for first place must be at least 3. We know there are at least two events in the track meet, and that Lincoln High (which won the shot-put) had a final score of 9, so the integer for first place cannot be more than 8. Can it be 8? No, because then only two events could take place and there is no way that Washington High could build up a total of 22 points. It cannot be 7 because this permits no more than three events, and three are still not sufficient to enable Washington High to reach a score of 22. Slightly more involved arguments eliminate 6, 4 and 3 as the integer for first place. Only 5 remains as a possibility.

If 5 is the value for first place, there must be at least five events in the meet. (Fewer events are not sufficient to give Washington a total of 22, and more than five would raise Lincoln's total to more than 9.) Lincoln scored 5 for the shot-put, so its four other scores must be 1. Washington can now reach 22 in only two ways: 4, 5, 5, 5, 3 or 2, 5, 5, 5, 5. The first is eliminated because it gives Roosevelt a score of 17, and we know that this score is 9. The remaining possibility gives Roosevelt a correct final tally, so we have the unique reconstruction of the scoring shown in the table.

EVENTS	1	2	3	4	5	SCORE
WASHINGTON	2	5	5	5	5	22
LINCOLN	5	1	1	1	1	9
ROOSEVELT	1	2	2	2	2	9

Washington High won all events except the shot-put, consequently it must have won the high jump.

Many readers sent shorter solutions than the one just given. Two readers (Mrs. Erlys Jedlicka, Saratoga, California, and Albert Zoch, a student at Illinois Institute of Technology) noticed that there was a short cut to the solution based on the assumption that the problem had a unique answer. Mrs. Jedlicka put it this way:

> *Dear Mr. Gardner:*
>
> *Did you know this problem can be solved without any calculation whatever? The necessary clue is in the last paragraph. The solution to the integer equations must indicate without ambiguity which school won the high jump. This can only be done if one school has won all the events, not counting the shot-put; otherwise the problem could not be solved with the information given, even after calculating the scoring and number of events. Since the school that won the shot-put was not the over-all winner, it is obvious that the over-all winner won the remaining events. Hence without calculation it can be said that Washington High won the high jump.*

47. It is not possible for the termite to pass once through the 26 outside cubes and end its journey in the center one. This is easily demonstrated by imagining that the cubes alternate in color like the cells of a three-dimensional checkerboard, or the sodium and chlorine atoms in the cubical crystal lattice of ordinary salt. The large cube will then consist of 13 cubes of one color and 14 of the other color. The termite's path is always through cubes that alternate in color along the way; therefore if the path is to include all 27 cubes, it must begin and end with a cube belonging to the set of 14. The central cube, however, belongs to the 13 set; hence the desired path is impossible.

The problem can be generalized as follows: A cube of even order (an even number of cells on the side) has the same number of cells of one color as it has cells of the other color. There is no central cube, but complete paths may start on any cell and end on any cell of opposite color. A cube of odd order has one more cell of one color than the other, so a complete path must begin and end on the color that is used for the larger set. In odd-order cubes of orders 3, 7, 11, 15, 19 . . . the central cell belongs to the smaller set, so it cannot be the end of any complete path. In odd-order cubes of 1, 5, 9, 13, 17 . . . the central cell belongs to the larger set, so it can be the end of any path that starts on a cell of the same color. No closed path, going through every unit cube, is possible on any odd-order cube because of the extra cube of one color.

Many two-dimensional puzzles can be solved quickly by similar "parity checks." For example, it is not possible for a rook to start at one corner of a chessboard and follow a path that carries it once through every square and ends on the square at the diagonally opposite corner.

48. The dime and penny puzzle can be solved in four moves as follows. Coins are numbered from left to right.

1. Move 3, 4 to the right of 5 but separated from 5 by a gap equal to the width of two coins.
2. Move 1, 2 to the right of 3, 4, with coins 4 and 1 touching.
3. Move 4, 1 to the gap between 5 and 3.
4. Move 5, 4 to the gap between 3 and 2.

49. I originally published a solution showing how the three slices of bread could be toasted in two minutes. Five readers surprised me by cutting the time to 111 seconds. I had overlooked the possibility of partially toasting one side of a slice, removing it, then returning it later to complete the toasting.

Seconds	*Operation*
1–3	Put in slice A.
3–6	Put in B.
6–18	A completes 15 seconds of toasting on one side.
18–21	Remove A.
21–23	Put in C.
23–36	B completes toasting on one side.
36–39	Remove B.
39–42	Put in A, turned.
42–54	Butter B.
54–57	Remove C.
57–60	Put in B.
60–72	Butter C.
72–75	Remove A.
75–78	Put in C.
78–90	Butter A.
90–93	Remove B.
93–96	Put in A, turned to complete the toasting on its partially toasted side.
96–108	A completes its toasting.
108–111	Remove C.

All slices are now toasted and buttered, but slice A is still in the toaster. Even if A must be removed to complete the entire operation, the time is only 114 seconds.

50. A man goes up a mountain one day, down it another day. Is there a spot along the path that he occupies at the same time of day on both trips? This problem was called to my attention by psychologist Ray Hyman, of the University of Oregon, who in turn found it in a monograph entitled "On Problem-Solving," by the German Gestalt psychologist Karl Duncker. Duncker writes of being unable to solve it and of observing with satisfaction that others to whom he put the problem had the same difficulty. There are several ways to go about it, he continues, "but probably none is . . . more drastically evident than the following. Let ascent and descent be divided between *two* persons on the same day. They must *meet.* Ergo. . . . With this, from an unclear dim condition not easily surveyable, the situation has suddenly been brought into full daylight."

51. Immanuel Kant calculated the exact time of his arrival home as follows. He had wound his clock before leaving, so a glance at its face told him the amount of time that had elapsed during his absence. From this he subtracted the length of time spent with Schmidt (having checked Schmidt's hallway clock when he arrived and again when he left). This gave him the total time spent in walking. Since he returned along the same route, at the same speed, he halved the total walking time to obtain the length of time it took him to walk home. This added to the time of his departure from Schmidt's house gave him the time of his arrival home.

52. The first step is to list in order the probability values for the nine cards: 1/45, 2/45, 3/45. . . . The two lowest values are combined to form a new element: 1/45 plus 2/45 equals 3/45. In other words, the probability that the chosen card is either an ace or deuce is 3/45. There are now eight elements: the ace-deuce set, the three, the four, and so on up to nine. Again the two lowest probabilities are combined: the ace-deuce value of 3/45 and the 3/45 probability that the card is a three. This new element, consisting of aces, deuces and threes, has a probability value of 6/45. This is greater than the values for either the fours or fives, so when the two lowest values are combined again, we must pair the fours and fives to obtain an element with the value of 9/45. This procedure of pairing the lowest elements is continued until only one element remains. It will have the probability value of 45/45, or 1.

This chart shows how the elements are combined. The strategy for minimizing the number of questions is to take these pairings in reverse order. Thus the first question could be: Is the card in the set of fours, fives and nines? If not, you know it is in the other set so you ask next: Is it a seven or eight? And so on until the card is guessed.

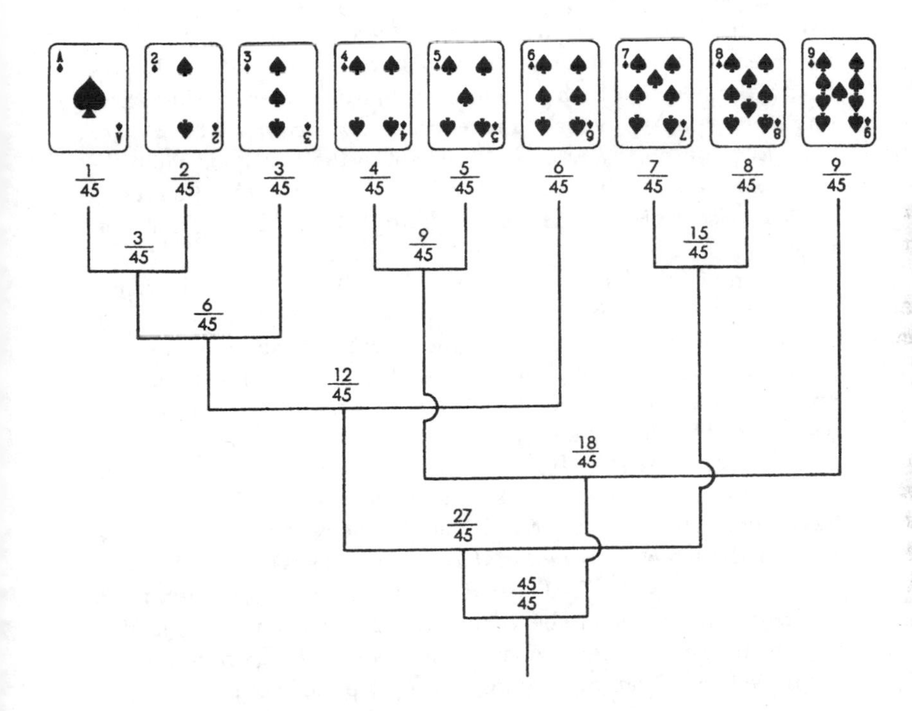

Note that if the card should be an ace or deuce it will take five questions to pinpoint it. A binary strategy, of simply dividing the elements as nearly as possible into halves for each question, will ensure that no more than four questions need be asked, and you might even guess the card in three. Nevertheless, the previously described procedure will give a slightly lower expected minimum number of questions in the long run; in fact, the lowest possible. In this case, the minimum number is three.

The minimum is computed as follows: Five questions are needed if the card is an ace. Five are also needed if the card is a deuce, but there are two deuces, making ten questions in all. Similarly, the three threes call for three times four, or 12, questions. The total number of questions for all 45 cards is 135, or an average of three questions per card.

This strategy was first discovered by David A. Huffman, an electrical engineer at M.I.T., while he was a graduate student there. It is explained in his paper "A Method for the Construction of Minimum-Redundancy Codes," *Proceedings of the Institute of Radio Engineers,* Vol. 40, pages 1098–1101, September 1952. It was later rediscovered by Seth Zimmerman, who described it in his article on "An Optimal Search Procedure," *American Mathematical Monthly,* Vol. 66, pages 690–693, October 1959. A good nontechnical exposition of the procedure will be found in John R. Pierce, *Symbols, Signals and Noise* (Harper & Brothers, 1961), beginning on page 94.

The eminent mathematician Stanislaw Ulam, in his biography *Adventures of a Mathematician* (Scribner's, 1976, page 281), suggested adding the following rule to the Twenty Questions game. The person who answers is permitted one lie. What is the minimum number of questions required to determine a number between 1 and one million? What if he lies *twice?*

The general case is far from solved. If there are no lies, the answer is of course 20. If just one lies, 25 questions suffice. This was proved by Andrzej Pelc, in "Solution of Ulam's Problem on Searching with a Lie," in *Journal of Combinatorial Theory* (Series A), Vol. 44, pages 129–140, January 1987. The author also gives an algorithm for finding the minimum number of needed questions for identifying any number between 1 and n. A different proof of the 25 minimum is given by Ivan Niven in "Coding Theory Applied to a Problem of Ulam," *Mathematics Magazine,* Vol. 61, pages 275–281, December 1988.

When two lies are allowed, the answer of 29 questions was established by Jurek Czyzowicz, Andrzej Pelc and Daniel Mundici in the *Journal of Combinatorial Theory* (Series A), Vol. 49, pages 384–388, November 1988. In the same journal (Vol. 52, pages 62–76, September 1989), the same authors solved the more general case of two lies and any number between 1 and 2^n. Wojciech Guziki, *ibid.,* Vol. 54, pages 1–19, 1990, completely disposed of the two-lie case for any number between 1 and n.

How about *three* lies? This has been answered only for numbers between 1 and one million. The solution is given by Alberto Negro and Matteo Sereno, in the same journal, Vol. 59, 1992. It is 33 questions, and that's no lie.

The four-lie case remains unsolved even for numbers in the 1 to one million range. Of course if one is allowed to lie every time, there is no way to guess the number. Ulam's problem is closely related to error-correcting coding theory. Ian Stewart summarized the latest results in "How to Play Twenty Questions with a Liar," in *New Scientist* (October 17, 1992), and Barry Cipra did the same in "All Theorems Great and Small," *SIAM News* (July 1992, page 28).

53. In the chess problem white can avoid checkmating black only by moving his rook four squares to the west. This checks the black king, but black is now free to capture the checking bishop with his rook.

When this problem appeared in *Scientific American,* dozens of readers complained that the position shown is not a possible one because there are two white bishops on the same color squares. They forgot that a pawn on the last row can be exchanged for any piece, not just the queen. Either of the two missing white pawns could have been promoted to a second bishop.

There have been many games by masters in which pawns were promoted to knights. Promotions to bishops are admittedly rare, yet one can imagine situations in which it would be desirable. For instance, to avoid stalemating the opponent. Or white may see that he can use either a new queen or a new bishop in a subtle checkmate. If he calls for a queen, it will be taken by a black rook, in turn captured by a white knight. But if white calls for a bishop, black may be reluctant to trade a rook for bishop and so let the bishop remain.

54. The seven varieties of convex hexahedrons, with topologically distinct skeletons, are shown in the illustration (next page). I know of no simple way to prove that there are no others. An informal proof is given by John McClellan in his article on "The Hexahedra Problem," *Recreational Mathematics Magazine,* No. 4, August 1961, pages 34–40.

There are 34 topologically distinct convex heptahedra, 257 octahedra and 2,606 9-hedra. There are three nonconvex (concave or re-entrant) hexahedra, 26 nonconvex heptahedra and 277 nonconvex octahedra. See the following papers by P. J. Federico: "Enumeration of Polyhedra: The Number of 9-Hedra," *Journal of Combinatorial*

Theory, Vol. 7, September 1969, pages 155–161; "Polyhedra with 4 to 8 Faces," *Geometria Dedicata,* Vol. 3, 1975, pages 469–481; and "The Number of Polyhedra," *Philips Research Reports,* Vol. 30, 1975, pages 220–231.

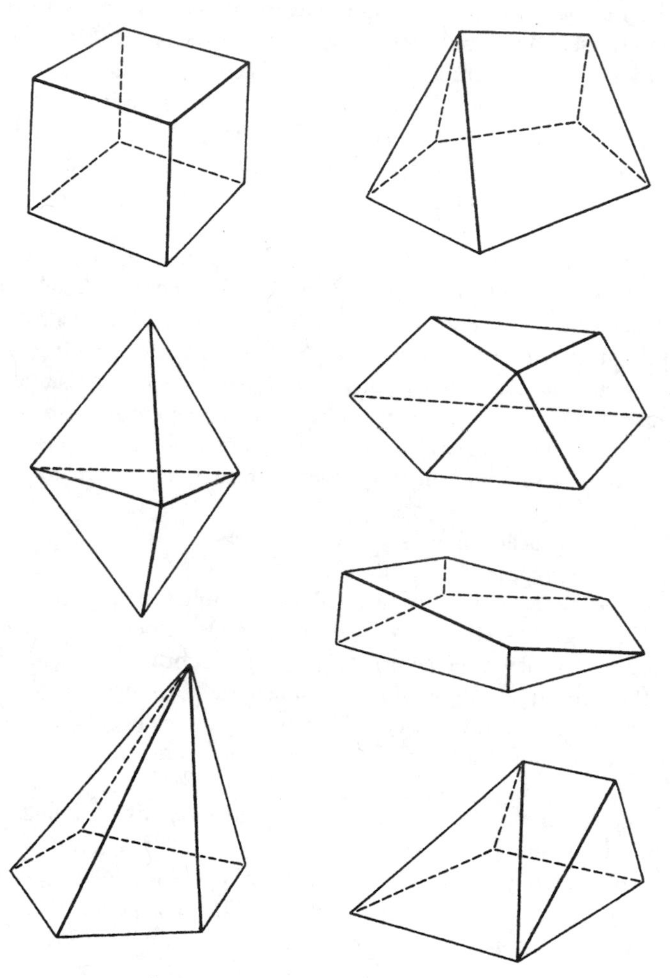

A formula for calculating the number of topologically distinct convex polyhedra, given the number of faces, remains undiscovered.

Paul R. Burnett called my attention to the Old Testament verse Zechariah 3:9. In a modern translation by J. M. Powis Smith it reads:

"For behold the stone which I have set before Joshua; upon a single stone with seven facets I will engrave its inscription."

An outline of a formal proof that there are just seven distinct convex hexahedra is given in "Euler's Formula for Polyhedra and Related Topics," by Donald Crowe, in *Excursions into Mathematics*, by Anatole Beck, Michael Bleicher and Donald Crowe (Worth, 1969, pages 29–30).

55.

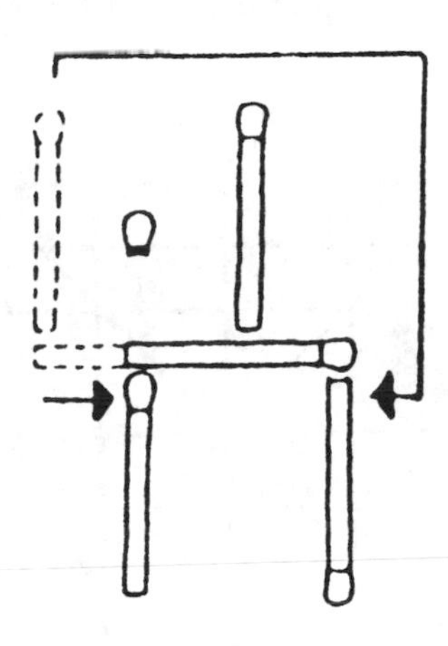

56. It is only necessary to cut the three links of one piece. They can then be used to join the remaining three pieces into the circular bracelet.

57.

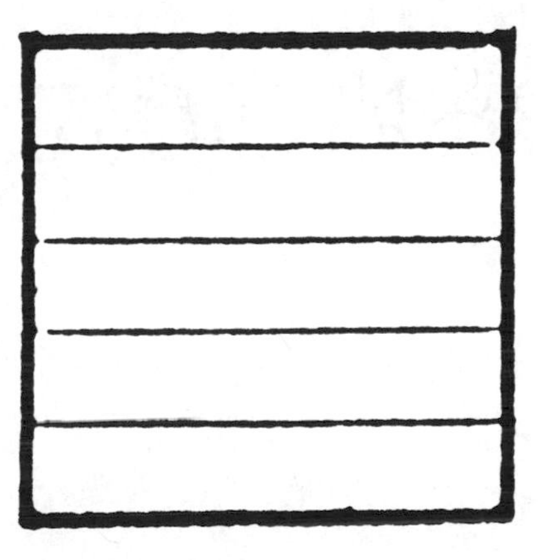

58. Continue the deal by taking cards from the *bottom* of the packet of undealt cards, dealing first to yourself, then counter-clockwise around the table.

59. Sal wins again. In the first race she ran 100 yards in the time it took Saul to run 90. Therefore, in the second race, after Saul has gone 90 yards, Sal will have gone 100, so she will be alongside him. Both will have 10 more yards to go. Since Sal is the faster runner, she will finish before Saul.

60.

61. The symbols are the numerals 1, 2, 3, 4, 5, 6, 7 shown alongside their mirror reflections. The next symbol, therefore, is the double 8, as shown at the far right, below.

62. The figure is cut into congruent halves like this:

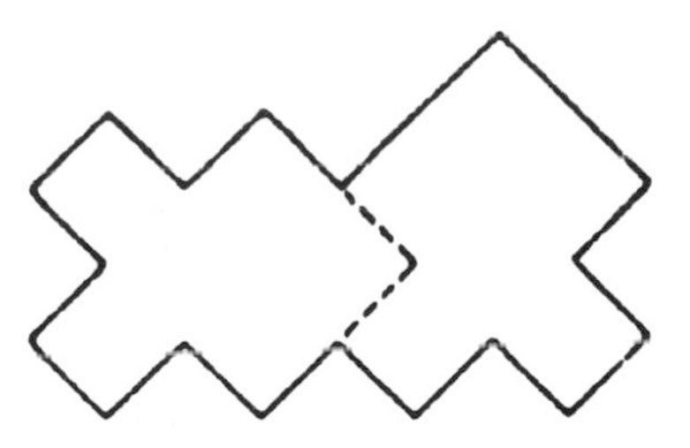

63. Two weighings will do the job. Divide the nine balls into three sets of triplets. Weigh one triplet against another. If a pan goes down you know the heavy ball is among the three on that pan. Pick any two of these balls and weigh one against the other. If one side goes down, you have found the ball. If they balance, the heavy ball must be the one you put aside. In either case, you have found the odd ball in two weighings.

Suppose the two triplets balance on the first weighing. You know then that the heavy ball is in the remaining triplet. As described above, the heavier ball of this triplet is easily identified by weighing any ball of the triplet against any other.

64. Cross out every other letter, starting with N. This eliminates NINE LETTERS, leaving A SINGLE WORD.

After this answer appeared in *Games,* Don Dwyer, Jr., sent to the magazine a second solution. Cross out the nine letters AEILNRSTW every time they appear, to leave the word GOD.

65.

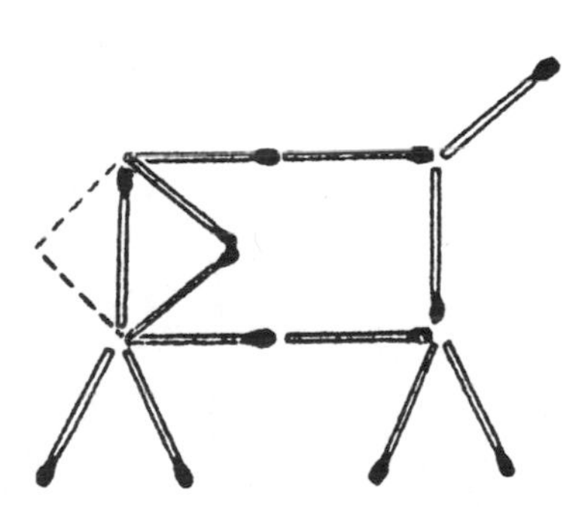

66. As the pictures below show, the folded letter is an upside-down and turned-over F.

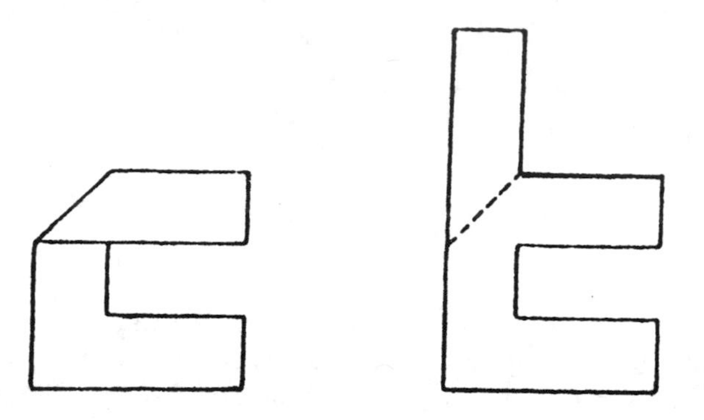

A CATALOG OF SELECTED

DOVER BOOKS

IN ALL FIELDS OF INTEREST

A CATALOG OF SELECTED DOVER BOOKS IN ALL FIELDS OF INTEREST

100 BEST-LOVED POEMS, Edited by Philip Smith. "The Passionate Shepherd to His Love," "Shall I compare thee to a summer's day?" "Death, be not proud," "The Raven," "The Road Not Taken," plus works by Blake, Wordsworth, Byron, Shelley, Keats, many others. 96pp. 5³⁄₁₆ x 8¼. 0-486-28553-7

100 SMALL HOUSES OF THE THIRTIES, Brown-Blodgett Company. Exterior photographs and floor plans for 100 charming structures. Illustrations of models accompanied by descriptions of interiors, color schemes, closet space, and other amenities. 200 illustrations. 112pp. 8⅜ x 11. 0-486-44131-8

1000 TURN-OF-THE-CENTURY HOUSES: With Illustrations and Floor Plans, Herbert C. Chivers. Reproduced from a rare edition, this showcase of homes ranges from cottages and bungalows to sprawling mansions. Each house is meticulously illustrated and accompanied by complete floor plans. 256pp. 9⅜ x 12¼. 0-486-45596-3

101 GREAT AMERICAN POEMS, Edited by The American Poetry & Literacy Project. Rich treasury of verse from the 19th and 20th centuries includes works by Edgar Allan Poe, Robert Frost, Walt Whitman, Langston Hughes, Emily Dickinson, T. S. Eliot, other notables. 96pp. 5³⁄₁₆ x 8¼. 0-486-40158-8

101 GREAT SAMURAI PRINTS, Utagawa Kuniyoshi. Kuniyoshi was a master of the warrior woodblock print — and these 18th-century illustrations represent the pinnacle of his craft. Full-color portraits of renowned Japanese samurais pulse with movement, passion, and remarkably fine detail. 112pp. 8⅜ x 11. 0-486-46523-3

ABC OF BALLET, Janet Grosser. Clearly worded, abundantly illustrated little guide defines basic ballet-related terms: arabesque, battement, pas de chat, relevé, sissonne, many others. Pronunciation guide included. Excellent primer. 48pp. 4³⁄₁₆ x 5¾. 0-486-40871-X

ACCESSORIES OF DRESS: An Illustrated Encyclopedia, Katherine Lester and Bess Viola Oerke. Illustrations of hats, veils, wigs, cravats, shawls, shoes, gloves, and other accessories enhance an engaging commentary that reveals the humor and charm of the many-sided story of accessorized apparel. 644 figures and 59 plates. 608pp. 6 ⅛ x 9¼. 0-486-43378-1

ADVENTURES OF HUCKLEBERRY FINN, Mark Twain. Join Huck and Jim as their boyhood adventures along the Mississippi River lead them into a world of excitement, danger, and self-discovery. Humorous narrative, lyrical descriptions of the Mississippi valley, and memorable characters. 224pp. 5³⁄₁₆ x 8¼. 0-486-28061-6

ALICE STARMORE'S BOOK OF FAIR ISLE KNITTING, Alice Starmore. A noted designer from the region of Scotland's Fair Isle explores the history and techniques of this distinctive, stranded-color knitting style and provides copious illustrated instructions for 14 original knitwear designs. 208pp. 8⅜ x 10⅞. 0-486-47218-3

Browse over 9,000 books at www.doverpublications.com

CATALOG OF DOVER BOOKS

ALICE'S ADVENTURES IN WONDERLAND, Lewis Carroll. Beloved classic about a little girl lost in a topsy-turvy land and her encounters with the White Rabbit, March Hare, Mad Hatter, Cheshire Cat, and other delightfully improbable characters. 42 illustrations by Sir John Tenniel. 96pp. 5³⁄₁₆ x 8¼. 0-486-27543-4

AMERICA'S LIGHTHOUSES: An Illustrated History, Francis Ross Holland. Profusely illustrated fact-filled survey of American lighthouses since 1716. Over 200 stations — East, Gulf, and West coasts, Great Lakes, Hawaii, Alaska, Puerto Rico, the Virgin Islands, and the Mississippi and St. Lawrence Rivers. 240pp. 8 x 10¾. 0-486-25576-X

AN ENCYCLOPEDIA OF THE VIOLIN, Alberto Bachmann. Translated by Frederick H. Martens. Introduction by Eugene Ysaye. First published in 1925, this renowned reference remains unsurpassed as a source of essential information, from construction and evolution to repertoire and technique. Includes a glossary and 73 illustrations. 496pp. 6⅛ x 9¼. 0-486-46618-3

ANIMALS: 1,419 Copyright-Free Illustrations of Mammals, Birds, Fish, Insects, etc., Selected by Jim Harter. Selected for its visual impact and ease of use, this outstanding collection of wood engravings presents over 1,000 species of animals in extremely lifelike poses. Includes mammals, birds, reptiles, amphibians, fish, insects, and other invertebrates. 284pp. 9 x 12. 0-486-23766-4

THE ANNALS, Tacitus. Translated by Alfred John Church and William Jackson Brodribb. This vital chronicle of Imperial Rome, written by the era's great historian, spans A.D. 14-68 and paints incisive psychological portraits of major figures, from Tiberius to Nero. 416pp. 5³⁄₁₆ x 8¼. 0-486-45236-0

ANTIGONE, Sophocles. Filled with passionate speeches and sensitive probing of moral and philosophical issues, this powerful and often-performed Greek drama reveals the grim fate that befalls the children of Oedipus. Footnotes. 64pp. 5³⁄₁₆ x 8 ¼. 0-486-27804-2

ART DECO DECORATIVE PATTERNS IN FULL COLOR, Christian Stoll. Reprinted from a rare 1910 portfolio, 160 sensuous and exotic images depict a breathtaking array of florals, geometrics, and abstracts — all elegant in their stark simplicity. 64pp. 8⅜ x 11. 0-486-44862-2

THE ARTHUR RACKHAM TREASURY: 86 Full-Color Illustrations, Arthur Rackham. Selected and Edited by Jeff A. Menges. A stunning treasury of 86 full-page plates span the famed English artist's career, from *Rip Van Winkle* (1905) to masterworks such as *Undine, A Midsummer Night's Dream,* and *Wind in the Willows* (1939). 96pp. 8⅜ x 11. 0-486-44685-9

THE AUTHENTIC GILBERT & SULLIVAN SONGBOOK, W. S. Gilbert and A. S. Sullivan. The most comprehensive collection available, this songbook includes selections from every one of Gilbert and Sullivan's light operas. Ninety-two numbers are presented uncut and unedited, and in their original keys. 410pp. 9 x 12. 0-486-23482-7

THE AWAKENING, Kate Chopin. First published in 1899, this controversial novel of a New Orleans wife's search for love outside a stifling marriage shocked readers. Today, it remains a first-rate narrative with superb characterization. New introductory Note. 128pp. 5³⁄₁₆ x 8¼. 0-486-27786-0

BASIC DRAWING, Louis Priscilla. Beginning with perspective, this commonsense manual progresses to the figure in movement, light and shade, anatomy, drapery, composition, trees and landscape, and outdoor sketching. Black-and-white illustrations throughout. 128pp. 8⅜ x 11. 0-486-45815-6

Browse over 9,000 books at www.doverpublications.com

THE BATTLES THAT CHANGED HISTORY, Fletcher Pratt. Historian profiles 16 crucial conflicts, ancient to modern, that changed the course of Western civilization. Gripping accounts of battles led by Alexander the Great, Joan of Arc, Ulysses S. Grant, other commanders. 27 maps. 352pp. 5⅜ x 8½. 0-486-41129-X

BEETHOVEN'S LETTERS, Ludwig van Beethoven. Edited by Dr. A. C. Kalischer. Features 457 letters to fellow musicians, friends, greats, patrons, and literary men. Reveals musical thoughts, quirks of personality, insights, and daily events. Includes 15 plates. 410pp. 5⅜ x 8½. 0-486-22769-3

BERNICE BOBS HER HAIR AND OTHER STORIES, F. Scott Fitzgerald. This brilliant anthology includes 6 of Fitzgerald's most popular stories: "The Diamond as Big as the Ritz," the title tale, "The Offshore Pirate," "The Ice Palace," "The Jelly Bean," and "May Day." 176pp. 5⅜ x 8½. 0-486-47049-0

BESLER'S BOOK OF FLOWERS AND PLANTS: 73 Full-Color Plates from Hortus Eystettensis, 1613, Basilius Besler. Here is a selection of magnificent plates from the *Hortus Eystettensis,* which vividly illustrated and identified the plants, flowers, and trees that thrived in the legendary German garden at Eichstätt. 80pp. 8⅜ x 11. 0-486-46005-3

THE BOOK OF KELLS, Edited by Blanche Cirker. Painstakingly reproduced from a rare facsimile edition, this volume contains full-page decorations, portraits, illustrations, plus a sampling of textual leaves with exquisite calligraphy and ornamentation. 32 full-color illustrations. 32pp. 9⅜ x 12¼. 0-486-24345-1

THE BOOK OF THE CROSSBOW: With an Additional Section on Catapults and Other Siege Engines, Ralph Payne-Gallwey. Fascinating study traces history and use of crossbow as military and sporting weapon, from Middle Ages to modern times. Also covers related weapons: balistas, catapults, Turkish bows, more. Over 240 illustrations. 400pp. 7¼ x 10⅞. 0-486-28720-3

THE BUNGALOW BOOK: Floor Plans and Photos of 112 Houses, 1910, Henry L. Wilson. Here are 112 of the most popular and economic blueprints of the early 20th century — plus an illustration or photograph of each completed house. A wonderful time capsule that still offers a wealth of valuable insights. 160pp. 8⅜ x 11. 0-486-45104-6

THE CALL OF THE WILD, Jack London. A classic novel of adventure, drawn from London's own experiences as a Klondike adventurer, relating the story of a heroic dog caught in the brutal life of the Alaska Gold Rush. Note. 64pp. 5³⁄₁₆ x 8¼. 0-486-26472-6

CANDIDE, Voltaire. Edited by Francois-Marie Arouet. One of the world's great satires since its first publication in 1759. Witty, caustic skewering of romance, science, philosophy, religion, government — nearly all human ideals and institutions. 112pp. 5³⁄₁₆ x 8¼. 0-486-26689-3

CELEBRATED IN THEIR TIME: Photographic Portraits from the George Grantham Bain Collection, Edited by Amy Pastan. With an Introduction by Michael Carlebach. Remarkable portrait gallery features 112 rare images of Albert Einstein, Charlie Chaplin, the Wright Brothers, Henry Ford, and other luminaries from the worlds of politics, art, entertainment, and industry. 128pp. 8⅜ x 11. 0-486-46754-6

CHARIOTS FOR APOLLO: The NASA History of Manned Lunar Spacecraft to 1969, Courtney G. Brooks, James M. Grimwood, and Loyd S. Swenson, Jr. This illustrated history by a trio of experts is the definitive reference on the Apollo spacecraft and lunar modules. It traces the vehicles' design, development, and operation in space. More than 100 photographs and illustrations. 576pp. 6¾ x 9¼. 0-486-46756-2

A CHRISTMAS CAROL, Charles Dickens. This engrossing tale relates Ebenezer Scrooge's ghostly journeys through Christmases past, present, and future and his ultimate transformation from a harsh and grasping old miser to a charitable and compassionate human being. 80pp. $5\frac{3}{16}$ x $8\frac{1}{4}$. 0-486-26865-9

COMMON SENSE, Thomas Paine. First published in January of 1776, this highly influential landmark document clearly and persuasively argued for American separation from Great Britain and paved the way for the Declaration of Independence. 64pp. $5\frac{3}{16}$ x $8\frac{1}{4}$. 0-486-29602-4

THE COMPLETE SHORT STORIES OF OSCAR WILDE, Oscar Wilde. Complete texts of "The Happy Prince and Other Tales," "A House of Pomegranates," "Lord Arthur Savile's Crime and Other Stories," "Poems in Prose," and "The Portrait of Mr. W. H." 208pp. $5\frac{3}{16}$ x $8\frac{1}{4}$. 0-486-45216-6

COMPLETE SONNETS, William Shakespeare. Over 150 exquisite poems deal with love, friendship, the tyranny of time, beauty's evanescence, death, and other themes in language of remarkable power, precision, and beauty. Glossary of archaic terms. 80pp. $5\frac{3}{16}$ x $8\frac{1}{4}$. 0-486-26686-9

THE COUNT OF MONTE CRISTO: Abridged Edition, Alexandre Dumas. Falsely accused of treason, Edmond Dantès is imprisoned in the bleak Chateau d'If. After a hair-raising escape, he launches an elaborate plot to extract a bitter revenge against those who betrayed him. 448pp. $5\frac{3}{16}$ x $8\frac{1}{4}$. 0-486-45643-9

CRAFTSMAN BUNGALOWS: Designs from the Pacific Northwest, Yoho & Merritt. This reprint of a rare catalog, showcasing the charming simplicity and cozy style of Craftsman bungalows, is filled with photos of completed homes, plus floor plans and estimated costs. An indispensable resource for architects, historians, and illustrators. 112pp. 10 x 7. 0-486-46875-5

CRAFTSMAN BUNGALOWS: 59 Homes from "The Craftsman," Edited by Gustav Stickley. Best and most attractive designs from Arts and Crafts Movement publication — 1903–1916 — includes sketches, photographs of homes, floor plans, descriptive text. 128pp. $8\frac{1}{4}$ x 11. 0-486-25829-7

CRIME AND PUNISHMENT, Fyodor Dostoyevsky. Translated by Constance Garnett. Supreme masterpiece tells the story of Raskolnikov, a student tormented by his own thoughts after he murders an old woman. Overwhelmed by guilt and terror, he confesses and goes to prison. 480pp. $5\frac{3}{16}$ x $8\frac{1}{4}$. 0-486-41587-2

THE DECLARATION OF INDEPENDENCE AND OTHER GREAT DOCUMENTS OF AMERICAN HISTORY: 1775-1865, Edited by John Grafton. Thirteen compelling and influential documents: Henry's "Give Me Liberty or Give Me Death," Declaration of Independence, The Constitution, Washington's First Inaugural Address, The Monroe Doctrine, The Emancipation Proclamation, Gettysburg Address, more. 64pp. $5\frac{3}{16}$ x $8\frac{1}{4}$. 0-486-41124-9

THE DESERT AND THE SOWN: Travels in Palestine and Syria, Gertrude Bell. "The female Lawrence of Arabia," Gertrude Bell wrote captivating, perceptive accounts of her travels in the Middle East. This intriguing narrative, accompanied by 160 photos, traces her 1905 sojourn in Lebanon, Syria, and Palestine. 368pp. $5\frac{3}{8}$ x $8\frac{1}{2}$. 0-486-46876-3

A DOLL'S HOUSE, Henrik Ibsen. Ibsen's best-known play displays his genius for realistic prose drama. An expression of women's rights, the play climaxes when the central character, Nora, rejects a smothering marriage and life in "a doll's house." 80pp. $5\frac{3}{16}$ x $8\frac{1}{4}$. 0-486-27062-9

Browse over 9,000 books at www.doverpublications.com